IT'S HERE!

PRENTICE HALL
SCIENCE

FINALLY, THE PERFECT FIT.

NOW YOU CAN CHOOSE THE PERFECT FIT FOR ALL YOUR CURRICULUM NEEDS.

The new Prentice Hall Science program consists of 19 hardcover books, each of which covers a particular area of science. All of the sciences are represented in the program so you can choose the perfect fit to *your* particular curriculum needs.

The flexibility of this program will allow you to teach those topics you want to teach, and to teach them *in-depth*. Virtually any approach to science—general, integrated, coordinated, thematic, etc.—is possible with Prentice Hall Science.

Above all, the program is designed to make your teaching experience easier and more fun.

ELECTRICITY AND MAGNETISM

Ch. 1. Electric Charges and Currents
Ch. 2. Magnetism
Ch. 3. Electromagnetism
Ch. 4. Electronics and Computers

HEREDITY: THE CODE OF LIFE

Ch. 1. What is Genetics?
Ch. 2. How Chromosomes Work
Ch. 3. Human Genetics
Ch. 4. Applied Genetics

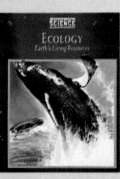

ECOLOGY: EARTH'S LIVING RESOURCES

Ch. 1. Interactions Among Living Things
Ch. 2. Cycles in Nature
Ch. 3. Exploring Earth's Biomes
Ch. 4. Wildlife Conservation

PARADE OF LIFE: MONERANS, PROTISTS, FUNGI, AND PLANTS

Ch. 1. Classification of Living Things
Ch. 2. Viruses and Monerans
Ch. 3. Protists
Ch. 4. Fungi
Ch. 5. Plants Without Seeds
Ch. 6. Plants With Seeds

EXPLORING THE UNIVERSE

Ch. 1. Stars and Galaxies
Ch. 2. The Solar System
Ch. 3. Earth and Its Moon

EVOLUTION: CHANGE OVER TIME

Ch. 1. Earth's History in Fossils
Ch. 2. Changes in Living Things Over Time
Ch. 3. The Path to Modern Humans

EXPLORING EARTH'S WEATHER

Ch. 1. What Is Weather?
Ch. 2. What Is Climate?
Ch. 3. Climate in the United States

THE NATURE OF SCIENCE

Ch. 1. What is Science?
Ch. 2. Measurement and the Sciences
Ch. 3. Tools and the Sciences

ECOLOGY: EARTH'S NATURAL RESOURCES
Ch. 1. Energy Resources
Ch. 2. Earth's Nonliving Resources
Ch. 3. Pollution
Ch. 4. Conserving Earth's Resources

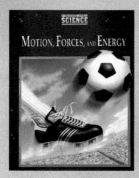

MOTION, FORCES, AND ENERGY
Ch. 1. What Is Motion?
Ch. 2. The Nature of Forces
Ch. 3. Forces in Fluids
Ch. 4. Work, Power, and Simple Machines
Ch. 5. Energy: Forms and Changes

PARADE OF LIFE: ANIMALS
Ch. 1. Sponges, Cnidarians, Worms, and Mollusks
Ch. 2. Arthropods and Echinoderms
Ch. 3. Fish and Amphibians
Ch. 4. Reptiles and Birds
Ch. 5. Mammals

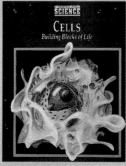

CELLS: BUILDING BLOCKS OF LIFE
Ch. 1. The Nature of LIfe
Ch. 2. Cell Structure and Function
Ch. 3. Cell Processes
Ch. 4. Cell Energy

DYNAMIC EARTH
Ch. 1. Movement of the Earth's Crust
Ch. 2. Earthquakes and Volcanoes
Ch. 3. Plate Tectonics
Ch. 4. Rocks and Minerals
Ch. 5. Weathering and Soil Formation
Ch. 6. Erosion and Deposition

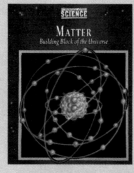

MATTER: BUILDING BLOCK OF THE UNIVERSE
Ch. 1. General Properties of Matter
Ch. 2. Physical and Chemical Changes
Ch. 3. Mixtures, Elements, and Compounds
Ch. 4. Atoms: Building Blocks of Matter
Ch. 5. Classification of Elements: The Periodic Table

CHEMISTRY OF MATTER
Ch. 1. Atoms and Bonding
Ch. 2. Chemical Reactions
Ch. 3. Families of Chemical Compounds
Ch. 4. Chemical Technology
Ch. 5. Radioactive Elements

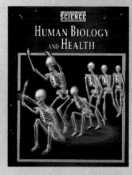

HUMAN BIOLOGY AND HEALTH
Ch. 1. The Human Body
Ch. 2. Skeletal and Muscular Systems
Ch. 3. Digestive System
Ch. 4. Circulatory System
Ch. 5. Respiratory and Excretory Systems
Ch. 6. Nervous and Endocrine Systems
Ch. 7. Reproduction and Development
Ch. 8. Immune System
Ch. 9. Alcohol, Tobacco, and Drugs

EXPLORING PLANET EARTH
Ch. 1. Earth's Atmosphere
Ch. 2. Earth's Oceans
Ch. 3. Earth's Fresh Water
Ch. 4. Earth's Landmasses
Ch. 5. Earth's Interior

HEAT ENERGY
Ch. 1. What Is Heat?
Ch. 2. Uses of Heat

SOUND AND LIGHT
Ch. 1. Characteristics of Waves
Ch. 2. Sound and Its Uses
Ch. 3. Light and the Electromagnetic Spectrum
Ch. 4. Light and Its Uses

A COMPLETELY INTEGRATED LEARNING SYSTEM...

The Prentice Hall Science program is an *integrated* learning system with a variety of print materials and multimedia components. All are designed to meet the needs of diverse learning styles and your technology needs.

THE STUDENT BOOK

Each book is a model of **excellent writing and dynamic visuals**— designed to be exciting and motivating to the student *and* the teacher, with relevant examples integrated throughout, and more opportunities for many different activities which apply to everyday life.

Problem-solving activities emphasize the thinking process, so problems may be more open-ended.

"Discovery Activities" throughout the book foster active learning.

Different sciences, and other disciplines, are integrated throughout the text and reinforced in the "Connections" features (the connections between computers and viruses is one example).

TEACHER'S RESOURCE PACKAGE

In addition to the student book, the complete teaching package contains:

ANNOTATED TEACHER'S EDITION

Designed to provide **"teacher-friendly"**

support regardless of instructional approach:

■ **Help is readily available** if you choose to teach thematically, to integrate the sciences, and/or to integrate the sciences with other curriculum areas.

■ **Activity-based learning** is easy to implement through the use of Discovery Strategies, Activity Suggestions, and Teacher Demonstrations.

■ Integration of all components is part of the teaching strategies.

■ For instant accessibility, all of the teaching sug-

gestions are wrapped around the student pages to which they refer.

ACTIVITY BOOK

Includes a **discovery activity for each chapter**, plus other activities including problem-solving and cooperative-learning activities.

THE REVIEW AND REINFORCEMENT GUIDE

Addresses **students' different learning styles** in a clear and comprehensive format:

■ Highly visual for visual learners.

TEACHER'S RESOURCE PACKAGE

FOR THE PERFECT FIT TO YOUR TEACHING NEEDS.

■ Can be used in conjunction with the program's audiotapes for auditory and language learners.

■ More than a study guide, it's a guide to comprehension, with activities, key concepts, and vocabulary.

ENGLISH AND SPANISH AUDIOTAPES
Correlate with the Review and Reinforcement Guide to aid auditory learners.

LABORATORY MANUAL ANNOTATED TEACHER'S EDITION
Offers **at least one additional hands-on opportunity per chapter** with

answers and teaching suggestions on lab preparation and safety.

TEST BOOK
Contains **traditional and up-to-the-minute strategies for student assessment.** Choose from performance-based tests in addition to traditional chapter tests and computer test bank questions.

STUDENT LABORATORY MANUAL
Each of the 19 books also comes with its own Student Lab Manual.

ALSO INCLUDED IN THE INTEGRATED LEARNING SYSTEM:

■ Teacher's Desk Reference

■ English Guide for Language Learners

■ Spanish Guide for Language Learners

■ Product Testing Activities

■ Transparencies

■ Computer Test Bank (IBM, Apple, or MAC)

■ VHS Videos

■ Videodiscs

■ Interactive Videodiscs (Level III)

■ Interactive Videodiscs/ CD ROM

■ Courseware

All components are integrated in the teaching strategies in the Annotated Teacher's Edition, where they directly relate to the science content.

THE PRENTICE HALL SCIENCE
INTEGRATED LEARNING SYSTEM

The following components are integrated in the teaching strategies for
EVOLUTION: CHANGE OVER TIME.

- Spanish Audiotape
  English Audiotape
- Activity Book
- Review and
  Reinforcement Guide
- Test Book—including
  Performance-Based Tests
- Laboratory Manual,
  Annotated Teacher's Edition
- Laboratory Manual

- English Guide for
  Language Learners
- Spanish Guide for
  Language Learners
- Transparencies:
  Half-Life
  Fossil Links
- Video/Videodisc:
  Patterns of Evolution

- Interactive Videodiscs:
  Insects: Little Giants of the
    Earth
  ScienceVision: TerraVision
- Interactive Videodiscs/
  CD ROM:
  Paul ParkRanger and
    the Mystery of the
    Disappearing Ducks
  Virtual BioPark
  Amazonia
  Science Discovery: Plate
    Tectonics

INTEGRATING OTHER SCIENCES

Many of the other 18 Prentice Hall Science books can be integrated into **EVOLUTION:**
CHANGE OVER TIME. The books you will find suggested most often in the Annotated Teacher's Edition
are DYNAMIC EARTH; EXPLORING PLANET EARTH; EXPLORING EARTH'S WEATHER;
CHEMISTRY OF MATTER; EXPLORING THE UNIVERSE; CELLS: BUILDING BLOCKS OF LIFE;
PARADE OF LIFE: MONERANS, PROTISTS, FUNGI, AND PLANTS; HEREDITY: THE CODE OF LIFE;
PARADE OF LIFE: ANIMALS; ECOLOGY: EARTH'S LIVING RESOURCES;
HUMAN BIOLOGY AND HEALTH; and SOUND AND LIGHT.

INTEGRATING THEMES

Many themes can be integrated into **EVOLUTION: CHANGE OVER TIME.**
Following are the ones most commonly suggested in the Annotated Teacher's Edition: EVOLUTION,
PATTERNS OF CHANGE, SYSTEMS AND INTERACTIONS, and UNITY AND DIVERSITY.

For more detailed information on teaching thematically and integrating
the sciences, see the Teacher's Desk Reference and teaching strategies throughout
the Annotated Teacher's Edition.

For more information, call 1-800-848-9500 or write:

 PRENTICE HALL

Simon & Schuster Education Group
113 Sylvan Avenue Route 9W
Englewood Cliffs, New Jersey 07632
Simon & Schuster A Paramount Communications Company

Annotated Teacher's Edition

Prentice Hall Science

Evolution
Change Over Time

Anthea Maton
Former NSTA National Coordinator
Project Scope, Sequence,
    Coordination
Washington, DC

Jean Hopkins
Science Instructor and Department
    Chairperson
John H. Wood Middle School
San Antonio, Texas

Susan Johnson
Professor of Biology
Ball State University
Muncie, Indiana

David LaHart
Senior Instructor
Florida Solar Energy Center
Cape Canaveral, Florida

Charles William McLaughlin
Science Instructor and Department
    Chairperson
Central High School
St. Joseph, Missouri

Maryanna Quon Warner
Science Instructor
Del Dios Middle School
Escondido, California

Jill D. Wright
Professor of Science Education
Director of International Field
    Programs
University of Pittsburgh
Pittsburgh, Pennsylvania

Prentice Hall
A Division of Simon & Schuster
Englewood Cliffs, New Jersey

ISBN 0-13-225533-2

2 3 4 5 6 7 8 9 10   97 96 95 94 93

Contents of Annotated Teacher's Edition

To the Teacher

Welcome to the *Prentice Hall Science* program. *Prentice Hall Science* has been designed as a complete program for use with middle school or junior high school science students. The program covers all relevant areas of science and has been developed with the flexibility to meet virtually all your curriculum needs. In addition, the program has been designed to better enable you—the classroom teacher—to integrate various disciplines of science into your daily lessons, as well as to enhance the thematic teaching of science.

The *Prentice Hall Science* program consists of nineteen books, each of which covers a particular topic area. The nineteen books in the *Prentice Hall Science* program are

The Nature of Science
Parade of Life: Monerans, Protists, Fungi, and Plants
Parade of Life: Animals
Cells: Building Blocks of Life
Heredity: The Code of Life
Evolution: Change Over Time

Ecology: Earth's Living Resources
Human Biology and Health
Exploring Planet Earth
Dynamic Earth
Exploring Earth's Weather
Ecology: Earth's Natural Resources
Exploring the Universe
Matter: Building Block of the Universe
Chemistry of Matter
Electricity and Magnetism
Heat Energy
Sound and Light
Motion, Forces, and Energy

Each of the student editions listed above also comes with a complete set of teaching materials and student ancillary materials. Furthermore, videos, interactive videos and science courseware are available for the *Prentice Hall Science* program. This combination of student texts and ancillaries, teacher materials, and multimedia products makes up your complete *Prentice Hall Science* Learning System.

About the Teacher's Desk Reference

The *Teacher's Desk Reference* provides you, the teacher, with an insight into the workings of the *Prentice Hall Science* program. The *Teacher's Desk Reference* accomplishes this task by including all the standard information you need to know about *Prentice Hall Science*.

The *Teacher's Desk Reference* presents an overview of the program, including a full description of each ancillary available in the program. It gives a brief summary of each of the student textbooks available in the *Prentice Hall Science* Learning System. The *Teacher's Desk Reference* also demonstrates how the seven science themes incorporated into *Prentice Hall Science* are woven throughout the entire program.

In addition, the *Teacher's Desk Reference* presents a detailed discussion of the features of the Student

Edition and the features of the Annotated Teacher's Edition, as well as an overview section that summarizes issues in science education and offers a message about teaching special students. Selected instructional essays in the *Teacher's Desk Reference* include English as a Second Language (ESL), Multicultural Teaching, Cooperative-Learning Strategies, and Integrated Science Teaching, in addition to other relevant topics. Further, a discussion of the Multimedia components that are part of *Prentice Hall Science*, as well as how they can be integrated with the textbooks, is included in the *Teacher's Desk Reference*.

The *Teacher's Desk Reference* also contains in blackline master form a booklet on Teaching Graphing Skills, which may be reproduced for student use.

Integrating the Sciences

The *Prentice Hall Science* Learning System has been designed to allow you to teach science from an integrated point of view. Great care has been taken to integrate other science disciplines, where appropriate, into the chapter content and visuals. In addition, the integration of other disciplines such as social studies and literature has been incorporated into each textbook.

On the reduced student pages throughout your Annotated Teacher's Edition you will find numbers within blue bullets beside selected passages and visuals. An Annotation Key in the wraparound margins indicates the particular branch of science or other discipline that has been integrated into the student text. In addition, where appropriate, the name of the textbook and the chapter number in which the particular topic is discussed in greater detail is provided. This enables you to further integrate a particular science topic by using the complete *Prentice Hall Science* Learning System.

Thematic Overview

When teaching any science topic, you may want to focus your lessons around the underlying themes that pertain to all areas of science. These underlying themes are the framework from which all science can be constructed and taught. The seven underlying themes incorporated into *Prentice Hall Science* are

Energy
Evolution
Patterns of Change
Scale and Structure
Systems and Interactions
Unity and Diversity
Stability

The primary themes in this textbook are Evolution, Patterns of Change, Systems and Interactions, and Unity and Diversity. Primary themes throughout *Prentice Hall Science* are denoted by an asterisk.

A detailed discussion of each of these themes and how they are incorporated into the *Prentice Hall Science* program are included in your *Teacher's Desk Reference*. In addition, the *Teacher's Desk Reference* includes thematic matrices for the *Prentice Hall Science* program.

A thematic matrix for each chapter in this textbook follows. Each thematic matrix is designed with the list of themes along the left-hand column and in the right-hand column a big idea, or overarching concept statement, as to how that particular theme is taught in the chapter.

CHAPTER 1

Earth's History in Fossils

ENERGY	
*EVOLUTION	• Fossils provide clear evidence that living things have changed, or evolved, over time.
*PATTERNS OF CHANGE	• Earth's landmasses have drifted apart from the supercontinent Pangaea, resulting in today's continents. • Earth's geologic history shows a series of changes in living things, climate, and land formations.
SCALE AND STRUCTURE	• Scientists use sedimentary rock layers to determine the relative ages of rocks and fossils. Fossils in lower layers are older than fossils in upper layers. • Absolute age can be determined using radioactive dating.
*SYSTEMS AND INTERACTIONS	• The remains of ancient organisms were often trapped in sediments. As the sediments hardened to form sedimentary rock, fossils formed.
*UNITY AND DIVERSITY	• Earth's geologic history and fossil record show that many varied organisms have evolved and become extinct during Earth's 4.6-billion-year history.
STABILITY	• The half-life of a radioactive element is fixed and cannot be changed.

CHAPTER 2

Changes in Living Things Over Time

ENERGY	
***EVOLUTION**	• Evolution is a change in species over time. • All living things have evolved from other living things.
***PATTERNS OF CHANGE**	• Mutations are the driving force behind evolution. Mutations that increase an organism's chances for survival are called adaptations. • Evolution may be gradual, as described by Darwin, or rapid, as described by Gould and Eldridge.
SCALE AND STRUCTURE	• Lamarck developed a theory of evolution based on homologous structures. • The molecular clock demonstrates how closely related two species are and when they branched off from a common ancestor.
***SYSTEMS AND INTERACTIONS**	• When two species share the same niche, they compete, often leading to the extinction of one of the species. • Adaptive radiation is the process by which one species evolves into several species, each of which fills a different niche.
***UNITY AND DIVERSITY**	• All members of a species share similar characteristics and can interbreed; they do exhibit minor variations from one another.
STABILITY	• Natural selection is the survival and reproduction of those organisms best adapted to their environment.

CHAPTER 3

The Path to Modern Humans

ENERGY	• Primates obtain their energy by eating plants and other animals.
*EVOLUTION	• The first human ancestors, which appeared about 6 million years ago, evolved from early primates.
*PATTERNS OF CHANGE	• As primates evolved, they developed certain characteristics that enabled them to survive in their environment.
SCALE AND STRUCTURE	• The body structures of primates vary according to their life functions.
*SYSTEMS AND INTERACTIONS	• Primates respond to and interact with their environment in ways that help them gather food, reproduce, and protect themselves.
*UNITY AND DIVERSITY	• Although each group of primates has different structures, they all have certain basic characteristics in common.
STABILITY	• The various life functions of primates help to maintain a stable internal and external environment.

Comprehensive List of Laboratory Materials

Item	Quantities per Group	Chapter
Adding machine tape	5 m	2
Meterstick	1 per student	2
Metric ruler	1	3
Milk carton, small	1	1
Petroleum jelly	1 jar	1
Plaster of Paris	1 box	1
Protractor	1 per student	3
Scissors	1 per student	3
Stirrer, metal rod or spoon	1	1

EVOLUTION
Change Over Time

Anthea Maton
Former NSTA National Coordinator
Project Scope, Sequence, Coordination
Washington, DC

Jean Hopkins
Science Instructor and Department Chairperson
John H. Wood Middle School
San Antonio, Texas

Charles William McLaughlin
Science Instructor and Department Chairperson
Central High School
St. Joseph, Missouri

Susan Johnson
Professor of Biology
Ball State University
Muncie, Indiana

Maryanna Quon Warner
Science Instructor
Del Dios Middle School
Escondido, California

David LaHart
Senior Instructor
Florida Solar Energy Center
Cape Canaveral, Florida

Jill D. Wright
Professor of Science Education
Director of International Field Programs
University of Pittsburgh
Pittsburgh, Pennsylvania

 Prentice Hall
Englewood Cliffs, New Jersey
Needham, Massachusetts

Prentice Hall Science

Evolution: Change Over Time

Student Text and Annotated Teacher's Edition
Laboratory Manual
Teacher's Resource Package
Teacher's Desk Reference
Computer Test Bank
Teaching Transparencies
Product Testing Activities
Computer Courseware
Video and Interactive Video

The illustration on the cover, rendered by Keith Kasnot, shows a modern iguana with the bones of an ancient dinosaur in the background.

Credits begin on page 117.

SECOND EDITION

© 1994, 1993 by Prentice-Hall, Inc., Englewood Cliffs, New Jersey 07632. All rights reserved. No part of this book may be reproduced in any form or by any means without permission in writing from the publisher. Printed in the United States of America.

ISBN 0-13-225525-1

2 3 4 5 6 7 8 9 10   97 96 95 94 93

Prentice Hall
A Division of Simon & Schuster
Englewood Cliffs, New Jersey 07632

STAFF CREDITS

Editorial:	Harry Bakalian, Pamela E. Hirschfeld, Maureen Grassi, Robert P. Letendre, Elisa Mui Eiger, Lorraine Smith-Phelan, Christine A. Caputo
Design:	AnnMarie Roselli, Carmela Pereira, Susan Walrath, Leslie Osher, Art Soares
Production:	Suse F. Bell, Joan McCulley, Elizabeth Torjussen, Christina Burghard
Photo Research:	Libby Forsyth, Emily Rose, Martha Conway
Publishing Technology:	Andrew Grey Bommarito, Deborah Jones, Monduane Harris, Michael Colucci, Gregory Myers, Cleasta Wilburn
Marketing:	Andrew Socha, Victoria Willows
Pre-Press Production:	Laura Sanderson, Kathryn Dix, Denise Herckenrath
Manufacturing:	Rhett Conklin, Gertrude Szyferblatt

Consultants

Kathy French	National Science Consultant
Jeannie Dennard	National Science Consultant
Brenda Underwood	National Science Consultant
Janelle Conarton	National Science Consultant

CONTENTS

EVOLUTION: CHANGE OVER TIME

Activity Bank/Reference Section

Features

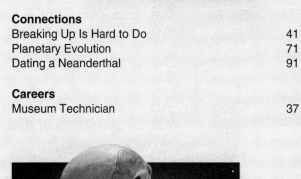

CONCEPT MAPPING

Throughout your study of science, you will learn a variety of terms, facts, figures, and concepts. Each new topic you encounter will provide its own collection of words and ideas—which, at times, you may think seem endless. But each of the ideas within a particular topic is related in some way to the others. No concept in science is isolated. Thus it will help you to understand the topic if you see the whole picture; that is, the interconnectedness of all the individual terms and ideas. This is a much more effective and satisfying way of learning than memorizing separate facts.

Actually, this should be a rather familiar process for you. Although you may not think about it in this way, you analyze many of the elements in your daily life by looking for relationships or connections. For example, when you look at a collection of flowers, you may divide them into groups: roses, carnations, and daisies. You may then associate colors with these flowers: red, pink, and white. The general topic is flowers. The subtopic is types of flowers. And the colors are specific terms that describe flowers. A topic makes more sense and is more easily understood if you understand how it is broken down into individual ideas and how these ideas are related to one another and to the entire topic.

It is often helpful to organize information visually so that you can see how it all fits together. One technique for describing related ideas is called a **concept map**. In a concept map, an idea is represented by a word or phrase enclosed in a box. There are several ideas in any concept map. A connection between two ideas is made with a line. A word or two that describes the connection is written on or near the line. The general topic is located at the top of the map. That topic is then broken down into subtopics, or more specific ideas, by branching lines. The most specific topics are located at the bottom of the map.

To construct a concept map, first identify the important ideas or key terms in the chapter or section. Do not try to include too much information. Use your judgment as to what is really important. Write the general topic at the top of your map. Let's use an example to help illustrate this process. Suppose you decide that the key terms in a section you are reading are School, Living Things, Language Arts, Subtraction, Grammar, Mathematics, Experiments, Papers, Science, Addition, Novels. The general topic is School. Write and enclose this word in a box at the top of your map.

SCHOOL

Now choose the subtopics—Language Arts, Science, Mathematics. Figure out how they are related to the topic. Add these words to your map. Continue this procedure until you have included all the important ideas and terms. Then use lines to make the appropriate connections between ideas and terms. Don't forget to write a word or two on or near the connecting line to describe the nature of the connection.

Do not be concerned if you have to redraw your map (perhaps several times!) before you show all the important connections clearly. If, for example, you write papers for Science as well as for Language Arts, you may want to place these two subjects next to each other so that the lines do not overlap.

One more thing you should know about concept mapping: Concepts can be correctly mapped in many different ways. In fact, it is unlikely that any two people will draw identical concept maps for a complex topic. Thus there is no one correct concept map for any topic! Even though your concept map may not match those of your classmates, it will be correct as long as it shows the most important concepts and the clear relationships among them. Your concept map will also be correct if it has meaning to you and if it helps you understand the material you are reading. A concept map should be so clear that if some of the terms are erased, the missing terms could easily be filled in by following the logic of the concept map.

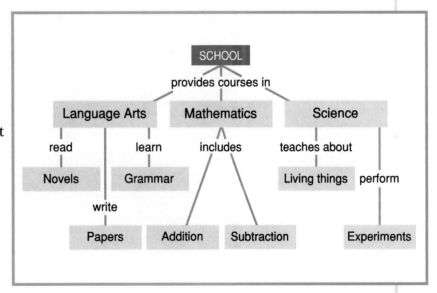

Evolution: Change Over Time

TEXT OVERVIEW

In this textbook students are introduced first to Earth's history. They learn about the kinds of fossils, about the dating of objects and events, and about geologic eras and periods. Next they study the subject of evolution. They are introduced to biochemical, anatomical, and fossil evidence of evolution. They learn about natural selection and the effects of overproduction, variation, migration, and isolation on evolutionary change.

Students learn about the general characteristics of the primates and then study characteristics that are unique to humans. They examine some of the fossil and chemical evidence that allows scientists to study human evolution, and they find out about some probable ancestors of humans. Finally, students look at the development of *Homo sapiens*.

TEXT OBJECTIVES

1. Identify six kinds of fossils and explain historical dating.
2. Name the eras and recent periods in geologic history and discuss events that took place in each.
3. Discuss the concept of evolution and the evidence that supports it.
4. Describe the chemical, anatomical, and fossil evidence of evolution.
5. Discuss some of the differences between modern humans and early primates.
6. Detail information about human evolution.

EVOLUTION
Change Over Time

▼ *Although neither the saber-toothed cat nor the ground sloth will survive its sticky encounter with the La Brea tar pits, its remains may become fossils found by scientists thousands of years in the future.*

A gentle breeze sent ripples across the shimmering surface of the shallow pond. The ground shook slightly as a giant ground sloth walked clumsily out of the nearby woods. Dipping its head toward the water, the sloth drank deeply. Suddenly, the sloth raised its head and stood very still, trying to detect a whisper of danger from the trees.

Hearing nothing, the sloth bent its head toward the water again. At that moment, a saber-toothed cat leaped from the woods. With

INTRODUCING EVOLUTION: CHANGE OVER TIME

USING THE TEXTBOOK

Begin your introduction of the textbook by having students examine the textbook-opening illustrations and captions. Before they read the textbook introduction, ask the following questions.

• **What do you see in the illustration on page F8?** (A saber-toothed cat jumping toward a ground sloth.)

• **Which animal is a predator and which is prey?** (The saber-toothed cat; the ground sloth.)

• **What do you see in the picture on page F9?** (An elephantlike creature.)

• **What happened to all these animals?** (They became engulfed by the La Brea tar pits.)

• **How do you think we know about these animals and what happened to them?** (Answers will vary, but guide students to understand that fossils of the animals were preserved in the pits and provide evi-

dence about them.)

• **Do these animals exist in the area of Los Angeles today?** Why or why not? (Accept all logical answers. Point out that the animals depicted are now extinct. The only information we have about them is from fossil evidence.)

Have students read the textbook introduction.

• **Are these the only fossils found?** (No. Fossils have been found in many other places.)

• **What information can fossils provide?**

1 Earth's History in Fossils In Chapter 1 six types of fossils are introduced. Fossils are defined not only as the remains of organisms but also as imprints and traces left by them. The ordering of events in natural history is explained. Finally, information on geologic areas and periods is provided.

2 Changes in Living Things Over Time Chapter 2 deals with the subject of evolution. Students are introduced to biochemical, anatomical, and fossil evidence of evolution. They learn about natural selection and the effects of overproduction, variation, migration, and isolation. Finally, they are introduced to a theory of large-scale, relatively rapid evolution.

3 The Path to Modern Humans In Chapter 3 students learn about the general characteristics of the primates and the characteristics unique to humans. They trace a probable lineage of development from early primates to humans as they learn that the Neanderthals and Cro-Magnons are closely related to modern humans.

claws bared and teeth flashing, the cat flew through the air toward its startled prey. With the giant cat on its back, the sloth plunged into the pond. Its feet splashed through the water but did not find solid ground. Instead, a sticky, gooey tar trapped the animal, pulling it further and further down into the pond.

In a matter of minutes, both animals were stuck in the tar. They remained there for thousands of years. In the early 1900s, scientists discovered their remains and pieced together a picture of the area in which the sloth and the cat had made their home. Today, this place is known as the La Brea tar pits in Los Angeles, California.

The types of fossils that were found in the La Brea pits are not the only kinds of fossils that have been found. As you read this book, you will discover some other types of evidence that help scientists gain an understanding of the Earth's past.

▲ *A model of an elephantlike creature trapped in tar can be seen at the La Brea tar pits in Los Angeles, California.*

Discovery *Activity*

Looking Into the Earth's Past

Take a walk through a nearby park or through your neighborhood and look for fossils embedded in the surfaces of sidewalks, building walls, rocks, and roads.

■ What do the fossils look like?
■ What do the fossils tell you about the kinds of living things that formed them?
■ How do you think the fossils formed?

(Fossils can provide information about the types of life forms that have inhabited Earth.)

• **Are the La Brea tar pits an important source of scientific information?** (Yes. The fossils found there have helped scientists piece together information about Earth's past.)

Point out that because the tar pits preserved the remains of the animals, scientists were able to piece together information about the creatures from Earth's past.

Ask students if there are other types of substances that may help to preserve remains. (Many students will recognize that ice is a preservative that has contained fossils.)

DISCOVERY ACTIVITY

Looking Into the Earth's Past

Begin the introduction to the textbook by having students perform the Discovery Activity. Help students identify fossils as more than the remains of living organisms. They should understand that fossils can be imprints of organisms as well as the organisms themselves. As students complete the activity, they should note that leaf imprints on the ground or footsteps in dried cement can be considered fossils. Explain that as students read the information in the chapters that follow, they will extend their understanding of fossils and the type of information that the fossils can provide about life in the past.

Chapter 1 | EARTH'S HISTORY IN FOSSILS

SECTION	HANDS-ON ACTIVITIES
1–1 Fossils — Clues to the Past pages F12–F18 Multicultural Opportunity 1–1, p. F12 ESL Strategy 1–1, p. F12	**Student Edition** ACTIVITY (Doing): Animal Footprints, p. F13 ACTIVITY (Discovering): Preservation in Ice, p. F14 LABORATORY INVESTIGATION: Interpreting Fossil Molds and Casts, p. F42 **Laboratory Manual** Interpreting Events From Fossil Evidence, p. F7 **Teacher Edition** How Fossils Form, p. F10d **Activity Book** CHAPTER DISCOVERY: Index Fossils, p. F9
1–2 A History in Rocks and Fossils pages F18–F28 Multicultural Opportunity 1–2, p. F18 ESL Strategy 1–2, p. F18	**Laboratory Manual** Exploring Geologic Time Through Sediment "Core" Samples, p. F11 Interpreting a Sediment Deposition Model, p. F15 Using the Rock Record to Interpret Geologic History, p. F21 Demonstrating Half-Life, p. F25 **Activity Book** ACTIVITY: Simulating Half-Life, p. F29 ACTIVITY: Constructing a Model of Radioactive Decay, p. F33
1–3 A Trip Through Geologic Time pages F29–F41 Multicultural Opportunity 1–3, p. F28 ESL Strategy 1–3, p. F28	**Student Edition** ACTIVITY (Doing): Pangaea, p. F36 ACTIVITY (Doing): The Time of Your Life, p. F39 **Teacher Edition** Units of Time, p. F10d
Chapter Review pages F42–F45	

OUTSIDE TEACHER RESOURCES

Books

A Field Guide to Dinosaurs. The Diagram Group, Avon.

Arnold, Caroline. *Trapped in Tar: Fossils From the Ice Age,* Clarion.

Branley, Franklyn M. *Dinosaurs, Asteroids, and Superstars: Why the Dinosaurs Disappeared,* Crowell Junior/Harper.

Clark, David L. *Fossils, Paleontology, and Evolution,* W. C. Brown.

LaPlante, Jerry C. *The Weekend Fossil Hunter,* Drake.

Sawkins, Frederick J., et al. *The Evolving Earth,* Macmillan.

Windley, B. F., ed. *The Early History of the Earth,* Wiley.

OTHER ACTIVITIES	MEDIA AND TECHNOLOGY
Student Edition ACTIVITY (Writing): Terrible Lizards, p. F15 **Activity Book** ACTIVITY: Fossil Types and Formations, p. F15 ACTIVITY: Fossil Facts, p. F17 ACTIVITY: Classifying Fossils, p. F21 **Review and Reinforcement Guide** Section 1–1, p. F5	**Transparency Binder** Fossil Links **English/Spanish Audiotapes** Section 1–1
Student Edition ACTIVITY (Writing): Index Fossils, p. F20 ACTIVITY (Calculating): Half-Life, p. F25 **Activity Book** ACTIVITY: Decay of Radioactive Material, p. F13 ACTIVITY: The Earth's Past, p. F25 ACTIVITY: Geologic Time Line, p. F43 **Review and Reinforcement Guide** Section 1–2, p. F9	**Transparency Binder** Half-Life **Courseware** Plate Tectonics (Supplemental) Continental Drift (Supplemental) **English/Spanish Audiotapes** Section 1–2
Student Edition ACTIVITY (Reading): Good Mother Lizard, p. F35 **Activity Book** ACTIVITY: Dinosaur Adaptations, p. F31 ACTIVITY: Life in the Geologic Time Scale, p. F37 ACTIVITY: Interpreting an Ancient Puzzle, p. F39 **Review and Reinforcement Guide** Section 1–3, p. F11	**Interactive Videodisc** ScienceVision: TerraVision **Interactive Videodisc/CD ROM** Science Discovery: Plate Tectonics **English/Spanish Audiotapes** Section 1–3
Test Book Chapter Test, p. F9 Performance-Based Tests, p. F71	**Test Book** Computer Test Bank Test, p. F15

*All materials in the Chapter Planning Guide Grid are available as part of the Prentice Hall Science Learning System.

Audiovisuals

Discovering Fossils, filmstrip with cassette, Encyclopaedia Britannica Education
Fossils, sound filmstrips, Encyclopaedia Britannica Education

Fossils: Clues to Earth History, study prints, Encyclopaedia Britannica Education
Fossils: Clues to the Past, video, National Geographic
Introduction to Fossils, Part 1 and 2, 40 slides, Society for Visual Education

Life Long Ago, filmstrips with cassettes, Society for Visual Education
Succession on Lava, 16-mm film, Encyclopaedia Britannica Education

CHAPTER OVERVIEW

Some of the best information about the history of the Earth has come from the study of fossils. Fossils may be the actual remains of a living thing or its traces. It is difficult to find complete fossil records because it is usually only the hard parts of a plant or animal that become fossilized.

Fossils may be in the form of molds, casts, imprints, or petrified remains. Sometimes entire organisms may be preserved in ice, amber, or tar.

Fossils reveal two main things about the Earth's history. First, they indicate that the surface of the Earth has changed over time. Second, they indicate that many different life forms have existed on the Earth during these times.

For the most part, fossils are found in sedimentary rocks, and these sediments are usually deposited in horizontal layers. The law of superposition provides evidence for these beliefs.

Geologists use the law of superposition to determine the relative age of rock layers and the fossils they contain. Index fossils, which lived only a limited period of geologic time, are also helpful in determining the relative age of rock layers.

1–1 FOSSILS—CLUES TO THE PAST
THEMATIC FOCUS

The purpose of this section is to introduce students to fossils and how they provide clues to the Earth's past. There are a number of different types of fossils, and each type is identified by the way it was formed. Fossils can be formed by petrification, as molds and casts when an organism becomes embedded in sediments, as an entire organism preserved in ice or amber, as imprints of objects trapped in mud or sand, and as trace fossils when an animal leaves tracks, tooth marks, or other signs of its activity.

The themes that can be focused on in this section are evolution and systems and interactions.

*Evolution: Fossils provide clear evidence that living things have changed, or evolved, over time.

*Systems and interactions: The remains of ancient organisms were often trapped in sediments. As the sediments hardened to form sedimentary rock, fossils formed.

PERFORMANCE OBJECTIVES 1–1

1. Explain how fossils provide information about the Earth's past.

2. Identify six different kinds of fossils.

SCIENCE TERMS 1–1

fossil p. F12

sediment p. F13

petrification p. F14

mold p. F14

cast p. F14

imprint p. F14

trace fossil p. F17

evolve p. F17

1–2 A HISTORY IN ROCKS AND FOSSILS
THEMATIC FOCUS

The purpose of this section is to introduce students to the methods used by scientists to determine the correct order of events in the Earth's history. Students will also learn how scientists are able to tell the approximate age of an object or event in the Earth's past.

One principle used in determining the age of rocks and fossils is the law of superposition. Special fossils called index fossils can also be used to identify the age of rock layers.

In addition, students will learn how faults, intrusions, and extrusions provide clues to the relative ages of rocks. Finally, the section explains how radioactive dating is used.

The themes that can be focused on in this section are scale and structure and stability.

Scale and structure: Scientists use sedimentary rock layers to determine the relative ages of rocks and fossils. Fossils in lower layers are older than fossils in upper layers.

Stability: The half-life of a radioactive element is fixed and cannot be changed.

PERFORMANCE OBJECTIVES 1–2

1. Describe how events in the Earth's history can be placed in the correct order.

2. Discuss the law of superposition.

3. Discuss the significance of index fossils.

4. Explain how faults, intrusions, and extrusions provide clues to the Earth's past.

5. Explain how radioactive dating is used to determine the ages of rocks and fossils.

SCIENCE TERMS 1–2

law of superposition p. F19

index fossil p. F21

unconformity p. F22

fault p. F22

intrusion p. F22

extrusion p. F23

half-life p. F25

1–3 A TRIP THROUGH GEOLOGIC TIME

THEMATIC FOCUS

The purpose of this section is to introduce students to the major divisions of geologic time—the era and the period. They will also be introduced to the four eras in chronological order: the Precambrian, the Paleozoic, the Mesozoic, and the Cenozoic.

Students will learn about the Earth's surface features and life forms that characterize each era. A two-page chart illustrating the geologic history of the Earth is included as part of the section.

The themes that can be focused on in this section are unity and diversity and patterns of change.

*Unity and diversity: Earth's geologic history and its fossil record show that many varied organisms have evolved and become extinct during the 4.6-billion-year history of the Earth.

*Patterns of change: Earth's geologic history shows a series of changes in living things, climate, and land formations.

PERFORMANCE OBJECTIVES 1–3

1. Identify the major divisions of geologic time.

2. Discuss the surface features and life forms that characterize each of the four eras of geologic time.

Discovery *Learning*

TEACHER DEMONSTRATIONS MODELING

How Fossils Form

The following demonstration will help students understand how fossils form in sedimentary rock. For the demonstration you will need sand of several different colors, several small dishes, a graduated cylinder, and several small objects, such as tiny sea shells or bits of twig or cork.

Place a different color sand in each dish. Mix some of the small objects into one or two of the dishes of sand. These objects will represent fossils. Now carefully pour the sand from each dish into the graduated cylinder creating distinct visible layers.

- **What do you notice about the different colors of sand?** (They have formed distinct layers.)
- **Where are the small objects that were mixed with the particular color of sand?** (The objects are found in the particular color of sand with which they were mixed.)
- **Can you tell by looking at the graduated cylinder which layer of sand was added first?** Explain. (Yes. The bottom layer was added first, the layer above it was added next, and so on.)

Units of Time

Take a sheet from a large month-by-month calendar and hold it up for the class to see. Tell students that the total area of the grid section of the sheet stands for a unit of time equal to one month. Use scissors to cut the grid into its component four (or five) horizontal strips, which correspond to weeks.

- **To what does each of the strips correspond?** (One week of time.)

Cut one of the strips into its seven component squares.

- **To what does each of these correspond?** (Each square corresponds to one day.)
- **How would I make a piece that corresponds to an hour?** To a minute? To a second? (Cut a small square into 24, then into 60, then into 60 again.)
- **How large a piece would result if this were possible?** (An extremely small piece would result.)

Point out that geologic time is also divided into units. Also point out that if the area of the original calendar grid represented the age of the Earth, the age of our human species would be represented by a piece about as small as the one that would represent one second when the paper's area stood for one month of time.

CHAPTER 1

Earth's History in Fossils

INTEGRATING SCIENCE

This life science chapter provides you with numerous opportunities to integrate other areas of science, as well as other disciplines, into your curriculum. Blue numbered annotations on the student page and integration notes on the teacher wraparound pages alert you to areas of possible integration.

In this chapter, you can integrate earth science and sedimentary rocks (p. 13), earth science and magma (p. 13), language arts (pp. 15, 20, 35, 39), earth science and climate (p. 17), earth science and sedimentation (p. 19), earth science and erosion and deposition (p. 19), earth science and faulting (p. 22), physical science and radioactivity (p. 25), earth science and Earth's moon (p. 27), social studies (p. 29), earth science and volcanoes (p. 32), life science and the origin of life (p. 32), earth science and Pangaea (p. 34), earth science and oceanography (p. 35), earth science and mass extinction (p. 37), earth science and ice age (p. 40), and earth science and continental drift (p. 41).

SCIENCE, TECHNOLOGY, AND SOCIETY/COOPERATIVE LEARNING

Test-tube dinosaurs? Not exactly, but some scientists think that recognizable "dinosaurs" could be produced using technology we have today. What's the recipe for a dinosaur? Dinosaur DNA, DNA replication technology, and fertile embryos of present-day reptiles are the ingredients scientists theorize they would need.

How could a viable sample of DNA be obtained from a group of organisms that have been extinct for millions of years? Insect fossils! A paleontologist from California suggested that genetic material from dinosaurs could be permanently preserved in insects that were fossilized by preservation in amber. Bloodsucking insects encased in amber could have undigested dinosaur blood in their stomachs. Insects in amber could represent genetic time capsules containing DNA from extinct species.

INTRODUCING CHAPTER 1

DISCOVERY LEARNING

▶ *Activity Book*

Begin your introduction to this chapter by using the Chapter 1 Discovery Activity from your *Activity Book*. Using this activity, students will discover how index fossils can be used to determine the relative ages of rock formations.

USING THE TEXTBOOK

Begin by having students read the chapter-opener text and observe the chapter-opener photograph.

• **What kinds of animals do you see in this drawing?** (Dinosaurs and other types of lizardlike creatures; the skeleton of some unknown animal in the foreground.)
• **What can you infer about the climate of the region shown?** (It appears to be tropical.)

Earth's History in Fossils

Guide for Reading

After you read the following sections, you will be able to

1–1 Fossils—Clues to the Past
- Describe how scientists use fossils as clues to events in Earth's past.

1–2 A History in Rocks and Fossils
- Define the law of superposition and describe how it is used to find the relative age of rocks and fossils.
- Describe how the half-life of radioactive elements is used to find the absolute age of rocks and fossils.

1–3 A Trip Through Geologic Time
- Describe the major life forms and geologic events that occurred in each geologic era.

A cloud of volcanic ash rises toward the sky above a strange landscape. Beneath the cloud, firs, pines, and tall palmlike trees sway in a warm, gentle breeze. Furry mammals, most no larger than a rat, scurry about on the forest floor. One day, descendants of these mammals will become the dominant animals on Earth. But that day is a long way off: The time is 140 million years ago, and the world belongs to the dinosaurs.

The world is changing, however. A great mountain chain is rising—a mountain chain that will eventually stretch from Alaska to Central America. Soon the landscape will be quite different. The Age of Dinosaurs will come to an end. All that will be left of these magnificent reptiles will be their bones.

One day in the distant future, a shepherd will build a cabin of dinosaur bones—the only such building in the world. Quite naturally, the place will be called Bone Cabin Quarry. And the vast area around it will be known as Wyoming!

In this chapter, you will learn about the Earth's past. You will also take a trip through time and discover many of the changes that have taken place in the Earth's 4.6-billion-year existence.

Journal *Activity*

You and Your World Although people and dinosaurs often do battle in Hollywood movies, all the dinosaurs were actually gone long before people evolved on Earth. But for now, imagine that you are alive during the Age of Dinosaurs. In your journal, write a brief story about a day in your life.

◀ *Slashing with their terrible sickle-shaped claws, a pack of fierce meat-eating dinosaurs attack an Iguanodon in this scene from 140 million years ago.*

F ■ 11

Once identified and extracted, dinosaur DNA could be manipulated by polymerase chain reaction. In this process even very small amounts of DNA replicate themselves into amounts great enough for experimentation. The dinosaur would then be introduced into present-day reptile embryos. The resulting offspring, theoretically, would be a recognizable version of an ancient dinosaur.

Cooperative learning: Using preassigned lab groups or randomly selected teams, have groups complete one of the following assignments.
- Write a segment for a weekly news magazine show on the possibility of "creating" modern-day dinosaurs. Their final product could be the script for their segment or, time permitting, the production of the segment for the class.
- Discuss the following questions.
1. Many people will react negatively to the possibility of producing dinosaurs from ancient DNA, but would there be positive outcomes?
2. What effect would these modern-day dinosaurs have on ecosystems?
Once groups have discussed these questions, classroom sharing of ideas and opinions could be used as a closure activity.

See Cooperative Learning in the *Teacher's Desk Reference.*

JOURNAL ACTIVITY

You may want to use the Journal Activity as the basis of a class discussion. Students might imagine themselves as an organism other than human; for example, an early mammal or even a plant. Students should be instructed to keep their Journal Activity in their portfolio.

- **Would you say that the area is sparsely or heavily populated with living things?** (Heavily populated.)
- **What is causing unusual conditions in the sky?** (An erupting volcano.)
Point out to students that many volcanoes still erupt on the Earth, but they were much more active in the North American West millions of years ago than they are today.
- **What event is described in the text that does not happen on the Earth today?** (The rising of a mountain chain.)

- **According to the text, what area of North America is shown in the drawing?** (Wyoming.)
- **How have the climate and landscape of Wyoming changed?** (Today Wyoming is drier and cooler; its landscape has less water and vegetation.)
- **How do scientists know what the climate of Wyoming was like 140 million years ago?** (Accept all reasonable answers.)

1–1 Fossils—Clues to the Past

Guide for Reading

Focus on these questions as you read.
▶ *What is a fossil?*
▶ *How do different kinds of fossils form?*

Figure 1–1 *Scientists carefully reassemble fossil bones to better understand the living things that existed long ago. What are fossils?*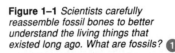

1–1 Fossils—Clues to the Past

The shepherd you have just read about used dinosaur bones to build Bone Cabin. Scientists use dinosaur bones, too, but for a different purpose. Dinosaur bones provide clues that help scientists build a different kind of structure—the structure of Earth's past. How exactly do they do this?

If you see a fish, you can conclude that somewhere on Earth there must be water, for that is where fishes live. If you see a polar bear, you can conclude that somewhere on Earth there is ice and cold temperatures—the environment in which polar bears live. In much the same way, scientists who study prehistoric forms of life use **fossils** to form a picture of Earth's past.

A fossil is the remains or evidence of a living thing. A fossil can be the bone of an organism or the print of a shell in a rock. A fossil can even be a burrow or tunnel left by an ancient worm. The most common fossils are bones, shells, pollen grains, and seeds.

Most fossils are not complete organisms. Fossils are generally incomplete because only the hard parts of dead plants or animals become fossils. The soft tissues either decay or are eaten before fossils can form. Decay is the breakdown of dead organisms into the substances from which they were made.

Most ancient forms of life have left behind few, if any, fossils as evidence that they once lived on Earth. In fact, the chances of any plant or animal leaving a fossil are slight at best. For most fossils to form, the remains of organisms usually have to be buried in **sediments** soon after the organisms die. Sediments are small pieces of rocks, shells, and other materials that were broken down over time. Quick burial in sediments prevents the dead organisms from being eaten by animals. It also slows down or stops the decay process.

Plants and animals that lived in or near water were preserved more often than other organisms were. Sediments in the form of mud and sand could easily bury plants and animals that died in the water or along the sides of a body of water. When the sediments slowly hardened and changed to sedimentary rocks, the organisms were trapped in the rocks. Sedimentary rocks are formed from layers of sediments. Most fossils are found in sedimentary rocks.

Rocks known as igneous rocks are formed by the cooling and hardening of hot molten rock, or magma. Most magma is found deep within the Earth, where no living things exist. Sometimes the magma flows onto the Earth's surface as hot, fiery lava. Can you explain why fossils are almost never found in igneous rocks? ❸

A third type of rock is called metamorphic rock. Metamorphic rocks are formed when sedimentary or igneous rocks are changed by heat, pressure, and chemical reactions. If there are fossils in a rock that undergoes such changes, the fossils are usually destroyed or damaged. So fossils are rarely found in metamorphic rocks as well.

There are many different kinds of fossils. Each kind is identified according to the process by which it was formed.

Petrification

When the dinosaurs died, the soft parts of their bodies quickly decayed. Only the hard parts—the bones—were left. Many of these bones were buried under layers of sediments of mud and wet sand. As water seeped through the layers of sediments, it dissolved minerals in the mud and sand. The water and

Figure 1–2 *These coiled shells once housed octopuslike animals that lived millions of years ago. How did the shells end up trapped in solid rock?* ❷

ACTIVITY DOING

Animal Footprints

1. Spread some mud in a low, flat-bottomed pan. Make sure the mud is not too wet and runny. Smooth the surface of the mud.

2. Have your pet or a neighbor's pet walk across the mud. Let the mud dry so that it hardens and the footprints are permanent.

3. Bring the footprints to science class. Exchange your set of footprints for the set of another student.

Examine the footprints and predict what type of animal made them. Explain how you arrived at your answer.

How is this activity similar to the way scientists determine what organism left the fossils that have been found?

F ■ 13

1–1 (continued)

CONTENT DEVELOPMENT

Describe petrification, drawing the attention of students to Figure 1–3, which shows petrified wood in the Petrified Forest National Park. Some students may have visited this or similar sites. If so, ask them to describe what they saw.

Go on to emphasize the similarities and differences between molds and casts. Point out that both form when a part of an organism becomes embedded in sand or mud that is turning into rock. In both cases the organic object eventually decays, leaving an empty space in the rock. If the space remains empty, the fossil is called a mold. If the space becomes filled with minerals, the fossil is called a cast.

REINFORCEMENT/RETEACHING

Review the differences between a mold and a cast, using the text and Figure 1–4 as a reference tool. Point out that molds can be filled in by scientists, using plaster. Such filled-in molds are called artificial casts to distinguish them from naturally formed casts. One such artificial cast was made of the remains of a rhinoceros that had been trapped in lava. The rear of the mold had been cut into and opened by the flowing action of the Columbia River. Scientists actually crawled inside the open mold in order to make a plaster cast of it.

Figure 1–3 *The Petrified Forest in Arizona has stone copies of trees that once grew there. The stone trees are the result of a process called petrification. How does petrification occur?* ①

ACTIVITY
DISCOVERING

Preservation in Ice

1. Place fresh fruit—such as apple slices, strawberries, and blueberries—in an open plastic container. Completely cover the fruit with water. Put the container in a freezer.

2. Put the same amount of fresh fruit in another open container. Place it somewhere where it will not be disturbed.

3. After three days, observe the fruit.

■ How do the samples in the two containers compare? How can you account for these differences?

minerals flowed through pores, or tiny holes, in the buried bones. When the water evaporated, the minerals were left behind in the bones, turning the bones to stone. This process is called **petrification,** which means turning into stone.

Petrification can occur in another way as well. Water may dissolve away animal or plant material. That material is replaced by the minerals in the water. This type of petrification is called replacement. Replacement produces an exact stone copy of the original animal or plant.

In the Petrified Forest of Arizona are some fossil trees that were created by replacement. Great stone logs up to 3 meters in diameter and more than 36 meters long lie in the Petrified Forest. Scientists suspect that the trees were knocked down by floods that swept over the land more than 200 million years ago. The remains of the trees—the fossil logs—show almost every detail of the once-living forest. For example, the patterns of growth rings in the trunks of many trees show up so clearly that scientists can count the growth rings and determine how long the trees lived.

Molds and Casts

Two types of fossils are formed when an animal or a plant is buried in sediments that harden into rock. If the soft parts of the organism decay and the hard parts are dissolved by chemicals, an empty space will be left in the rock. The empty space, called a **mold,** has the same shape as the organism.

Sometimes the mold is filled in by minerals in the sediment. The minerals harden to form a **cast,** or filled-in mold. The cast is in the same shape as the original organism.

Imprints

Sometimes a fossil is formed before the sediments harden into rock. Thin objects—such as leaves and feathers—leave **imprints,** or impressions, in soft sediments such as mud. When the sediments harden into rock, the imprints are preserved as fossils.

One particular imprint fossil has provided scientists with a clue to the development of the first

CONTENT DEVELOPMENT

Discuss the formation of imprint fossils. If you wish, make such a fossil by preparing a layer of wet clay-rich earth, pressing a leaf or a fern frond into it, removing the leaf or frond, and allowing the earth to harden.

Go on to explain freezing as a method of preserving organisms. Point out that bacteria, which account for decay, are unable to develop and reproduce at very low temperatures. As a result, the flesh of animals quickly frozen many thousands of

years ago may be edible if the animal is thawed. Dogs fed on the flesh of one of the first mammoths to be discovered. The mammoth had been frozen within a glacier.

REINFORCEMENT/RETEACHING

On the chalkboard, write the following terms: mold, cast, imprint, petrification, preservation of organism. Emphasize that these are five different types of fossils.

Figure 1–4 *Molds, casts, and imprints are types of fossils. Which part of the fossil shell is the mold? Which is the cast? The wing and tail feathers of this ancient bird are imprinted around its fossilized bones.*

birds. The imprint shows that the bird's skeleton was like that of a reptile with a toothed beak. Why do scientists believe the imprint was made by an ancient bird? The imprint also shows feathers around the skeleton, and only birds have feathers.

Preservation of Entire Organisms

Perhaps the most spectacular kinds of fossils are those in which the whole body, or complete sections of it, is preserved. Although it is quite rare for both the soft and the hard parts of an organism to be preserved, some entire-organism fossils exist. How was the decay of these organisms stopped completely so that they could be preserved?

FREEZING You probably know that freezing substances helps to preserve them. Freezing prevents substances from decaying. On occasion, scientists have found animals that have been preserved through freezing. Several extinct (no longer living on Earth) elephantlike animals called woolly mammoths have been discovered frozen in large blocks of ice. Woolly mammoths lived some 10,000 years ago. Another extinct animal, the furry rhinoceros, has been found preserved in the loose frozen soil in the arctic. So well preserved are the woolly mammoths and the furry rhinoceroses that wolves had sometimes eaten parts of the flesh when the ice had thawed.

ACTIVITY WRITING

Terrible Lizards

Using reference materials in the library, look up information about these reptiles:
Brachiosaurus
Ankylosaurus
Plesiosaurus
Write a report that includes a description of each reptile and its habitat. Accompany your description with a drawing.

F ■ 15

Also emphasize that fossils are classified according to how they are formed. Have students take turns describing the fossil types and giving examples of each.

ENRICHMENT

▶ *Activity Book*
Students who have mastered the concepts in this section will be challenged by the chapter activity Classifying Fossils.

GUIDED PRACTICE

▶ *Laboratory Manual*
Skills Development

Skills: Applying concepts, interpreting diagrams, making comparisons

At this point you may want to have students complete the Chapter 1 Laboratory Investigation in the *Laboratory Manual* called Interpreting Events From Fossil Evidence. In the investigation students will relate fossils to the history of an era.

Figure 1–5 *Fossils may form when living things are trapped in tree resin that later hardens into amber. The tiny scales of the lizard, the hairlike bristles on the cricket's hind legs, and the delicate wings of the termites are perfectly preserved.*

AMBER When the resin (sap) from certain evergreen trees hardens, it forms a hard substance called amber. Flies and other insects are sometimes trapped in the sticky resin that flows from these trees. When the resin hardens, the insects are preserved in the amber. Insects found trapped in amber are usually perfectly preserved.

TAR PITS Tar pits are large pools of tar. Tar pits contain the fossil remains of many different animals. The animals were trapped in the sticky tar when they went to drink the water that often covered the pits. Other animals came to feed on the trapped animals and were also trapped in the tar. Eventually, the trapped animals sank to the bottom of the tar pits. Bison, camels, giant ground sloths, wolves, vultures, and saber-toothed cats are some of the animals found as fossils in the tar pits. In the La Brea tar pits in present-day Los Angeles, California, the complete skeletons of animals that lived more than a million years ago are perfectly preserved.

Most of the fossils recovered from tar pits are bones. The flesh of the trapped animals had either decayed or been eaten before the animals could be preserved. But whole furry rhinoceroses have been found in tar pits in Poland.

Figure 1–6 *Dinosaur footprints are an example of trace fossils. What are some other types of trace fossils? Why are such fossils important to scientists?* ❶

16 ■ F

You may wish to have students do library research on tar preservation of organisms. Encourage students to prepare charts with drawings of such organisms and include with the names of the organisms, their approximate age, and the places in which they were discovered. An oral presentation can also be made to the class.

CONTENT DEVELOPMENT

Explain the meaning of the term extinction. Point out that fossil studies provide information on extinct organisms. Discuss the preservation in ice, amber, and tar of entire extinct organisms. Finally, explain the meaning and significance of trace fossils, and ask students to try to think of other examples of such fossils.

● ● ● ● **Integration** ● ● ● ●

Use the discussion of coral fossils to integrate concepts of climate into your lesson.

Trace Fossils

Trace fossils are fossils that reveal much about an animal's appearance without showing any part of the animal. Trace fossils are the marks or evidence of animal activities. Tracks, trails, footprints, and burrows are trace fossils. Trace fossils can be left behind by animals such as worms, which are too soft to be otherwise preserved.

Interpreting Fossils

Scientists can learn a great deal about Earth's past from fossils. **Fossils indicate that many different life forms have existed at different times throughout Earth's history.** In fact, some scientists believe that for every type of organism living today, there are at least 100 types of organisms that have become extinct.

When fossils are arranged according to age, they show that living things have **evolved**, or changed over time. By examining the changes in fossils of a particular type of living thing, scientists can determine how that living thing has evolved over many millions of years. You will learn a good deal more about the process of evolution in Chapter 2.

Fossils also indicate how the Earth's surface has evolved. For example, if scientists find fossils of sea organisms in rocks high above sea level, they can assume that the land was once covered by an ocean.

Fossils also give scientists clues to Earth's past climate. For example, fossils of coral have been found in arctic regions. Coral is an animal that lives only in warm ocean areas. So evidence of the presence of coral in arctic regions indicates that the climate in the Arctic was once much warmer than it is today. Fossils of alligators similar to those found in Florida today have been located as far north as Canada. What kind of climate might once have existed when these fossils were living? ❷

Fossils also tell scientists about the appearance and activities of extinct animals. From fossils of footprints, bones, and teeth, scientists construct models of extinct animals. They can even tell how big or heavy the animals were. Fossil footprints provide a clue as to how fast a particular animal could move.

Figure 1–7 *Scientists can learn much about the Earth's past from fossils. Although today's alligators live only in warm climates, alligatorlike fossils have been found as far north as Canada. What does this suggest about the past climate of Canada? What do the fossil shark teeth from a desert of Morocco indicate about that area's past? What characteristics of the shark's teeth indicate that it was a meat-eater?* ❸

F ■ 17

1–2 A History in Rocks and Fossils

MULTICULTURAL OPPORTUNITY 1–2

Have your students make a collection of legends and short stories from various cultures about the origin of the Earth. They might use the Native America Indian culture, the Egyptian culture, and the Arabic culture as three sources. Compare how these legends and stories explain the history of the Earth with how scientists explain Earth's history. In both the case of the cultural legends and the scientists, people are attempting to explain something that they cannot directly see. It will be helpful for your students to identify some of the ways in which the scientific approach is different from the approach of legends.

ESL STRATEGY 1–2

Have students read the following paragraph or read it to them. Discuss the paragraph with the class and then have students answer the question at the end.

You read that scientists have found evidence of a particular kind of organism in only one layer of rock near your home; later, the same organism is discovered in only one layer of rock in another part of the world. What will this discovery allow scientists to assume about the type and age of these fossils? (The fossils are probably of the same age.)

Point out the similarity in spelling of *intrusion* and *extrusion*. To help students understand the appropriateness of the geologic usage of these words, explain their Latin origins. *Intrusion* comes from a word meaning "*push in*," *extrusion* from "push out."

Although we may think of dinosaurs as being slow and plodding creatures, fossil footprints indicate that some dinosaurs could run as fast as 50 kilometers per hour. Fossils of teeth provide clues about the kind of food the animals ate. How might the shape of a tooth help scientists determine if an animal ate plants or other animals? ❶

1–1 Section Review

1. What is a fossil? List five different types and describe how each forms.
2. Do molds and casts represent the remains of organisms or evidence of those organisms? Explain.
3. How can fossils provide evidence of climate changes on Earth?

Critical Thinking—*Applying Concepts*
4. The Hawaiian Islands are volcanic in origin. Would you expect to find many fossils of ancient Hawaiian organisms on the islands?

Guide for Reading

Focus on these questions as you read.

▶ How do scientists use the law of superposition to determine the relative ages of rocks and fossils?

▶ How does the half-life of a radioactive element enable scientists to determine the absolute age of rocks and fossils?

1–2 A History in Rocks and Fossils

Using evidence from rocks and fossils, scientists can determine the order of events that occurred in the past: what happened first, second, third, and so on. And scientists can often approximate the time at which the events happened. In other words, scientists can "write" a history of Earth.

One way to think of Earth's history is to picture a very large book filled with many pages. Each page tells the story of an event in the past. The stories of the earliest events are in the beginning pages; the stories of later events are in the pages near the end.

Although you could tell from such a book which events occurred before others, you could not know when the events occurred. In other words, you could tell the order of the events, but not their dates. To know when an event occurred, you would need numbers on the pages. If each number stood for a

---

1–1 (continued)

INDEPENDENT PRACTICE

Section Review 1–1

1. Fossils are remains or traces of long-dead plants and animals. Petrified fossils are formed when minerals gradually replace the original substances of the plant or animal. Molds and casts are formed when an animal or plant is buried in sediments that harden into rock. Imprints are formed when thin objects leave an impression in soft sediment that eventually hardens. Preservation of whole animals takes place through freezing, amber, and tar pits.

2. Both molds and casts represent evidence because the original organism is no longer present in any form.

3. Fossils can provide evidence of climate changes because animals found in specific climates sometimes leave fossil remains in areas that have changed to another type of climate.

certain period of time—one million years perhaps—then you would have a fairly accurate calendar of Earth's history. As it turns out, scientists have both kinds of "history books" of Earth—one without page numbers but with a known order of pages, and one with page numbers as well.

The Law of Superposition

How, you might wonder, can scientists determine what events in Earth's history occurred before or after other events? That is, how do scientists develop their "book without dates"?

As you have read, most fossils are found in sedimentary rocks. Sedimentary rocks are made of layers of sediments that have piled up one atop the other. If the sediments have been left untouched, then clearly the layers of rocks at the bottom are older than those at the top. Put another way, the sedimentary layers are stacked in order of their age.

The **law of superposition** states that in a series of sedimentary rock layers, younger rocks normally lie on top of older rocks. The word superposition means one thing placed on top of another.

The law of superposition is based on the idea that sediments have been deposited in the same way throughout Earth's history. This idea was first proposed by the Scottish scientist James Hutton in the late eighteenth century. Hutton theorized that the processes acting on Earth's surface today are the same processes that have acted on Earth's surface in the past. These processes include weathering, erosion, and deposition. Weathering is the breaking down of rocks into sediments. Erosion is the carrying away of sediments. Deposition is the laying down of sediments.

Scientists use the law of superposition to determine whether a fossil or a layer of rock is older or younger than another fossil or layer of rock. Think of the layers of sedimentary rocks as the unnumbered pages of the "history book" that you have just read about. Remember, the beginning pages hold stories of long ago, while the end pages hold more recent stories. Now think of the words on each page as being a fossil. The words (fossil) on an earlier page (layer of rock) are older than the words (fossil) on a later page (layer of rock).

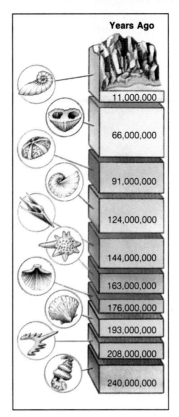

Years Ago

11,000,000
66,000,000
91,000,000
124,000,000
144,000,000
163,000,000
176,000,000
193,000,000
208,000,000
240,000,000

Figure 1–8 *Fossils are usually found in sedimentary rocks. If the sedimentary rock layers are in the same positions in which they formed, lower layers are older than upper ones. This principle is known as the law of superposition. How old are the fossils in the bottom layer of the diagram? The top layer?* ②

F ■ 19

filled with water. Next have them pour in a layer of small gravel, and then a layer of soil. After the water clears, the layers should be visible and in the sequence of "older" to "younger."

CONTENT DEVELOPMENT

Use the Focus/Motivation discussion to illustrate why older rock layers are usually found beneath younger rock layers in sedimentary rock. Emphasize that layers of sediments are laid down year after year. Unless subsurface activity disturbs the rock layers, the layers that were laid down first will remain on the bottom. Also emphasize that organic matter carried by sediments will remain with the sediments in which it was originally mixed. Thus, fossils found in the bottom layers of sedimentary rock will be older than fossils found in the top layers.

● ● ● ● **Integration** ● ● ● ●

Use the discussion of the law of superposition to integrate concepts of sedimentation into your lesson.

Use the discussion of Hutton's theories to integrate concepts of erosion and deposition into your lesson.

4. No, you would not expect to find many fossils of ancient Hawaiian organisms because volcanic lava melts living things that it traps. As a result, nothing will be left to form fossils.

REINFORCEMENT/RETEACHING

Review students' responses to the Section Review questions. Reteach any material that is still unclear, based on their responses.

CLOSURE

▶ *Review and Reinforcement Guide*
Have students complete Section 1–1 in the *Review and Reinforcement Guide*.

TEACHING STRATEGY 1–2

FOCUS/MOTIVATION

Have students simulate sedimentation to illustrate the law of superposition. To do this they should pour a layer of washed sand into an aquarium or 4-liter jar half

Figure 1–9 *The law of superposition helps scientists to determine the sequence of changes in life forms on the Earth. Is the spiral shell fossil older or younger than the cone-shaped shell fossil? How can you tell?* ●1

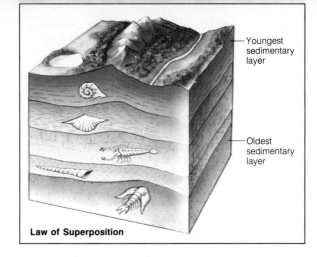

Youngest sedimentary layer

Oldest sedimentary layer

Law of Superposition

The process of sedimentation still occurs today. As you read these words, the fossils of tomorrow are being trapped in the sediments at the bottoms of rivers, lakes, and seas. For example, the Mississippi River deposits sediments at a rate of 80,000 tons an hour—day after day, year after year—at the point where the river flows into the Gulf of Mexico.

Index Fossils

The law of superposition helps scientists put in order the record of Earth's past for one particular location. But how can scientists get a worldwide picture of Earth's past? Is there a way of using the knowledge about the ages of rock layers in one location to find the relative ages of rock layers in other parts of the world? The relative age of an object is its age compared to the age of another object. Relative age does not provide dates for events, but it does provide a sequence of events.

In the early 1800s, scientists working on opposite sides of the English Channel came up with a way to determine relative ages of rock layers in different parts of the world. The scientists were digging through layers of sedimentary rocks near the southern coast of England and the northern coast of France. In both locations, the scientists discovered fossils of sea-dwelling shellfish. Clearly, both coasts had been under water at some time in the past.

1–2 (continued)

CONTENT DEVELOPMENT

Point out that scientists can use the law of superposition to determine the relative ages of organisms—or they can tell which organisms came first. To find the absolute age of an organism, scientists use techniques such as radioactive dating. Point out that students will learn more about radioactive dating at the end of Section 1–2.

CONTENT DEVELOPMENT

Have students examine Figure 1–9 and read the caption.
• **How many sedimentary layers are shown in the illustration?** (Five.)
• **Which layer is the oldest? The youngest?** (The bottom layer is the oldest; the top layer is the youngest.)
• **How many fossils are shown in the illustration?** (Five, one in each layer.)
• **What organisms left these fossils in the sediments?** (Accept all logical answers. Students may suggest simple invertebrates.)

FOCUS/MOTIVATION

Have students use reference materials to find out more information about index fossils. Have them make a list of index fossils. Encourage them to find out what geologists consider to be the most useful index fossils.

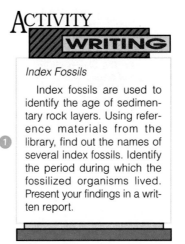

ACTIVITY
WRITING

Index Fossils

●1 Index fossils are used to identify the age of sedimentary rock layers. Using reference materials from the library, find out the names of several index fossils. Identify the period during which the fossilized organisms lived. Present your findings in a written report.

ENRICHMENT

Students may enjoy solving the following puzzle about relative age.
• **Darin is older than Nancy, who is older than John. John is younger than David, who is Nancy's younger brother. Jeffrey and Melinda are both older than Darin, but Melinda is not as old as Jeffrey. Can you rank the six people in order of age, from oldest to youngest?** (Jeffrey, Melinda, Darin, Nancy, David, John.)

CONTENT DEVELOPMENT

Emphasize that index fossils are used to determine the relative age of the rock layer in which the index fossil is found.
• **Suppose two rock layers in different locations contain fossils of the dinosaur *Triceratops*, which lived only during the Cretaceous Period. What must be true about these rock layers?** (They must have been formed at approximatley the same time, probably during the Cretaceous Period.)

As you can see from Figure 1–10, four distinct kinds of shellfish fossils were found on both sides of the channel. Fossil 1 was found only in the upper layers. Fossil 2 was found in various layers. Fossil 3 was found only in middle layers. And fossil 4 was found only in the lower layers. Because the same fossils were found in similar rock layers on both sides of the channel, the scientists concluded that layers with the same fossils were the same age. Thus fossils 1, 3, and 4 were clues to the relative ages of the rock layers. Fossil 2 was not. Explain why. ❷

Fossils such as fossils 1, 3, and 4 in Figure 1–10 are called **index fossils.** Index fossils are fossils of organisms that lived during only one short period of time. Scientists assume that index fossils of the same type of organism are all nearly the same age. So a layer of rock with one type of index fossil in it is close in age to another layer of rock with the same type of index fossil in it. Even though the rock layers may be in different regions of the world, the index fossils indicate that the layers are close in age. Why would the fossil of an organism that lived for a long period of time not be a good index fossil? ❸

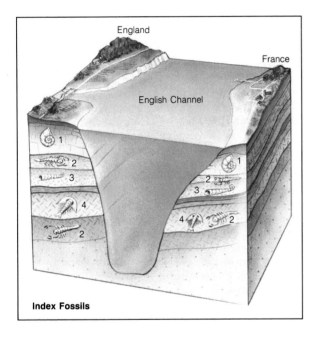

Index Fossils

Figure 1–10 *Index fossils are fossils of organisms that lived during only one short period of time. This illustration shows the rock layers on both sides of the English Channel. Even though the rock layers are separated by about 30 kilometers, there are three kinds of index fossils that are found on both sides. So scientists concluded that the English rock layers were the same age as the French rock layers that contain the same index fossils. Which three fossils helped scientists to reach this conclusion? Which fossil cannot be used to determine the relative age of the English and French rock layers?* ❹

F ■ 21

ANNOTATION KEY

Answers

❶ The spiral shell fossil is older because it is in a lower layer of rock. (Making inferences)

❷ It is found in more than one rock layer. It lived for too long a period of time to be useful. (Interpreting diagrams)

❸ It would not be possible to tell which period of time the fossil was from, so the fossil would be of no help in determining the age of rock layers. (Making inferences)

❹ Numbers 1, 3, and 4; number 2. (Interpreting diagrams)

Integration

❶ Language Arts

- **The trilobite called** *Olenellus* **lived only during the early Cambrian Period. Would you expect a rock layer containing this fossil to be found above or below a rock layer containing a fossil of the** *Triceratops?* **Explain.** (It would be found below the layer containing the *Triceratops* because the *Olenellus* lived during a much earlier time.)

- **Suppose a fossil of an organism that lived during the same times as the** *Triceratops* **and the** *Olenellus* **is found. Would this fossil be useful as an index fossil? Explain your answer.** (No. The fossil lived for too long a span of time. It would be impossible to determine the relative age of a rock layer containing this fossil.)

REINFORCEMENT/RETEACHING

Discuss the concept of index fossils using Figure 1–10 as your reference. Make sure students understand that the rock layers containing index fossils are often widely separated and that this technique can provide scientists with solid data about the relative ages of various organisms.

INDEPENDENT PRACTICE

▶ *Activity Book*

Students who need practice with the concepts of this section should be provided with the chapter activity The Earth's Past.

GUIDED PRACTICE

▶ *Laboratory Manual*

Skills Development

Skills: Making observations, applying concepts, making comparisons, making inferences

At this point you may want to have students complete the Chapter 1 Laboratory Investigation in the *Laboratory Manual* called Exploring Geologic Time Through Sediment "Core" Samples. In the investigation students will create and interpret sediment deposits.

GEOLOGIC FAULTS

There are a number of different kinds of geologic faults. These include normal faults, reverse faults, thrust faults, and lateral faults. In a normal fault the hanging wall, or block of rock lying above the fault, moves downward relative to the foot wall, or block of rock below the fault. In a reverse fault the hanging wall rises relative to the foot wall. In a thrust fault the hanging wall slides over the foot wall. In a lateral fault blocks slide horizontally past each other without rising or falling.

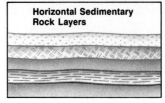
Horizontal Sedimentary Rock Layers

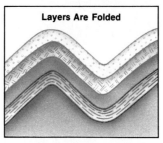
Layers Are Folded

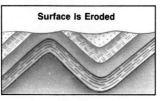
Surface Is Eroded

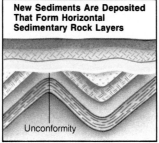
New Sediments Are Deposited That Form Horizontal Sedimentary Rock Layers

Unconformity

Figure 1–11 *One type of unconformity forms when forces within the Earth fold and tilt previously horizontal sedimentary rock layers (top and top center). In time, the layers are worn down to almost a flat surface (bottom center). After a long period of time, new rock layers form, covering the old surface and producing an unconformity (bottom).*

Unconformities

Sedimentary rock layers and the fossils found within them may be disturbed by powerful forces within the Earth. The rock layers may be folded, bent, and twisted. Sometimes older, deeply buried layers of rocks are uplifted to the Earth's surface. At the surface, the exposed rocks are weathered and eroded. Sediments are then deposited on top of the eroded surface of the older rocks. The deposited sediments harden to form new horizontal sedimentary rock layers. The old eroded surface beneath the newer rock layers is called an **unconformity.** Tilted sedimentary rock layers covered by younger horizontal sedimentary layers is an example of an unconformity. See Figure 1–11.

There is a wide gap in the ages of the rock layers above and below an unconformity. There is also a wide gap in the ages of the fossils in these rock layers. By studying unconformities, scientists can tell where and when the Earth's crust has undergone changes such as tilting, uplifting, and erosion. In addition, scientists can learn about the effects of these changes on the organisms living at that time.

Faults, Intrusions, and Extrusions

There are other clues to the relative ages of rocks and the history of events on Earth. During movements of the Earth's crust, rocks may break or crack. A break or crack along which rocks move is called a **fault.** The rock layers on one side of a fault may shift up or down relative to the rock layers on the other side of the fault.

Because faults can occur only after rock layers are formed, rock layers are always older than the faults they contain. The relative age of a fault can be determined from the relative age of the sedimentary layer that the fault cuts across. Scientists can determine the forces that have changed the Earth's surface by examining the faults in rock layers.

The relative ages of igneous rock formations can also be determined. Magma often forces its way into layers of rocks. The magma hardens in the rock layers and forms an **intrusion.** An intrusion is younger than the sedimentary rock layers it passes through.

1–2 (continued)

CONTENT DEVELOPMENT

Describe unconformities, making reference to the diagram in Figure 1–11, which illustrates this formation.

Explain to students that an unconformity occurs when older sedimentary rock is pushed up to the Earth's surface. After this upward movement has taken place, the older rock occupies a higher position than it would normally, based on its age.

After the older rock is worn down by weathering and erosion, new sediments are deposited on top of it. The unconformity is the border between the old, weathered rock and the new sediments. The rock formed from these new sediments is much younger than the older rock that was pushed upward. Thus, there is a wide gap in the ages of rock layers above and below an unconformity.

Point out that unconformities present some problems with respect to the law of superposition but that they also provide valuable information on changes in the Earth's crust and the forces that account for them.

● ● ● ● **Integration** ● ● ● ●

Use the discussion of movements of the Earth's crust to integrate concepts of faulting into your lesson.

ENRICHMENT

Layers of rock are said to be conformable where they have been deposited without interruption. Because of the opposing forces of uplift and erosion, however, there is no place on the Earth that has a complete conformable rock strata. The breaks that occur in the rocks are called unconformities. Angular unconformities are formed when horizontal layers are tilted upward and become leveled by erosion. Then younger horizontal layers are formed above them.

CONTENT DEVELOPMENT

Point out to students that just as intruding rock is always younger than the rock it pushes into, a fault is always younger than the rocks it disturbs. Because of this, scientists can determine the relative age of a fault according to the age of the surrounding rock.

FOCUS/MOTIVATION

Provide students with modeling clay of various colors, and have them construct models that illustrate unconformities,

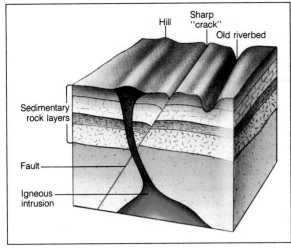

Sometimes magma reaches the surface of the Earth as lava and hardens. Igneous rock that forms on the Earth's surface is called an **extrusion.** Extrusions are younger than the rock layers beneath them. What do extrusions tell scientists about the Earth's past? ❶

A History With Dates

By the end of the nineteenth century, scientists had developed a clear picture of Earth's past as recorded by fossils and rocks. The picture showed great and varied changes. The continents had changed shape many times. High mountains had risen and had been worn away to hills. Life had begun in the sea and had later moved onto land. Living things had evolved through many stages to the forms that exist today. Climates around the world had also changed many times.

Scientists knew that these changes had taken place. They knew the order in which the changes had happened. But what they did not know was how many years the changes had taken. How many years ago had each event happened? It seemed clear that all the changes had taken a lot of time—certainly millions of years. But could a clock made of rock layers measure such lengths of time?

F ■ 23

faults, intrusions, and extrusions. Students can mount cross sections of these models on poster board to prepare a labeled chart for classroom display.

REINFORCEMENT/RETEACHING

Review the terms *intrusion* and *extrusion.* Remind students that scientists often use the absolute and relative ages of rock layers to determine the ages of fossils found in those rock layers.

CONTENT DEVELOPMENT

Have students examine Figure 1–12 and read the caption.
• **What effect has the fault had on the layers of sedimentary rock?** (The rock layers to the right have moved up with respect to the layers on the left.)
• **What effect has the intrusion had on the layers of rock?** (The intrusion is a break or interruption in the layers.)

HISTORICAL NOTE
CLOCKS

Before the invention of the pendulum clock, many clocks based on natural processes were used by ancient people. Some of these include the water clock, which measured time according to the rate of dripping water; the candle clock, which measured time according to the rate of burning wax; and the hourglass, which measured time according to the rate of falling sand.

ENRICHMENT

Have advanced students do library research on geologic formations in their own area and on fossils discovered there. They may wish to prepare a report and also a map of the area, with detailed information.

GUIDED PRACTICE
▶ *Laboratory Manual*
Skills Development

Skills: Making observations, making inferences, applying concepts

At this point you may want to have students complete the Chapter 1 Laboratory Investigation in the *Laboratory Manual* called Interpreting a Sediment Deposition Model. In the investigation, students will create a model of sediment deposition.

ENRICHMENT
▶ *Activity Book*

Students who have mastered the concepts in this section will be challenged by the chapter activity Geologic Time Line.

USING RADIOACTIVE ISOTOPES

Only certain types of rocks, mainly igneous, can be dated directly by atomic methods. When igneous rocks such as granite and basalt crystallize, the newly formed minerals contain various amounts of chemical elements, some of which have radioactive isotopes. These isotopes decay within the rocks according to their half-life rates. By selecting the appropriate minerals (those containing potassium, for example) and measuring the relative amounts of radioactive elements and decay elements, the date at which the rock crystallized can be determined.

Most sedimentary rocks such as sandstone, shale, and limestone are capable of being dated. This is done by comparing their ages with the ages of igneous rock formations that have been determined by radioactive dating.

1–2 (continued)

CONTENT DEVELOPMENT

Discuss with students the different types of "clocks" that geologists use to measure time. Point out that one type of clock is the sedimentary clock. The sedimentary clock is based on the assumption that sediments have been deposited at a steady rate throughout the Earth's history. Students can visualize this idea if they think of sand running through an hourglass. By noting the depth of a sedimentary rock layer, scientists can estimate the time at which it was laid down, and thus its age. The problem with this method of timekeeping is that rates of deposition have varied considerably at different times—so the sedimentary clock is not always accurate.

A much more precise method of geologic timekeeping is the radioactive clock. Stress that the rate of decay for a radioactive element is absolutely steady. By measuring the amount of a radioactive substance and the amount of its decay substance found in rocks or fossils, scientists can tell exactly how long the object has been in existence.

Figure 1–13 *The formation and wearing away of sedimentary rock layers do not occur at steady rates throughout geologic history. These rocks on the coast of New Zealand may have been built faster and worn away slower—or vice versa—than they were elsewhere. And the rates at which the rocks were built and destroyed may have varied greatly over time.*

To create a clock to measure time, you need to have a series of events that take place at a steady rate, like the steady movement of the second hand on a watch. The first clock developed by scientists to measure Earth's age was the rate at which sedimentary rock is deposited.

The scientists decided to assume that sediment was deposited at a steady rate throughout Earth's past. That rate, they reasoned, should be the same as it is in the present. Let's say the rate is 30 millimeters per century (100 years). If the total depth of sedimentary rocks deposited since the depositing began was measured and then divided by the yearly rate, the age of the oldest sedimentary rock could be calculated. In fact, this method could be used to determine the age of any sedimentary rock layer—and the fossils it contained.

In 1899, a British scientist used this method and came up with a maximum age for sedimentary rocks of about 75 million years. Although in some parts of the world this dating system seemed to work fairly well, it was not actually an accurate method of measuring time. Its greatest drawback was that there is not, and never has been, such a thing as a steady rate of deposition. A flood of the Mississippi River can lay down two meters of muddy sediment in a single day. But hundreds of years may pass before one meter of mud piles up at the bottom of a lake or pond.

So using the deposition of sediments to determine the age of Earth is like trying to use a clock that runs at widely different rates. Nevertheless, the invention of this "sedimentary clock" marked the beginning of efforts to measure the absolute age of events on Earth. Absolute age gives the precise time an event occurred, not just the order of events that relative age provides.

By the beginning of the twentieth century scientists who study the history and structure of the Earth agreed that Earth was at least several hundred million years old. Today, scientists know Earth's age to be about 4.6 billion years. What kind of clock do scientists use to measure the Earth's age so accurately? Strange as it may seem, the clock they use is a radioactive clock!

GUIDED PRACTICE

Skills Development

Skill: Interpreting photographs

Have students observe Figure 1–14. Before they read the caption, ask the following.

• **What is shown in the photograph?** (The skeleton of an animal.)

• **What type of animal do you think this skeleton is from?** (Accept all logical responses.)

Now have a volunteer read the caption.

• **What similarities do you see between the dinosaur skeleton and a human skeleton.** (Accept all logical responses. Students may point out the skull, ribs, arm, and hand.)

• **How would a scientist use radioactive dating to estimate how long ago the hatching dinosaur lived?** (Students should know that either the skeleton or the surrounding rock must have a radioactive element and an accompanying decay element in order to use radioactive dating.)

Radioactive Dating

The discovery of radioactive elements in 1896 led to the development of an accurate method of determining the absolute age of rocks and fossils. An atom of a radioactive element has an unstable nucleus, or center, that breaks down, or decays. During radioactive decay, particles and energy called radiation are released by the radioactive element.

As some radioactive elements decay, they form decay elements. A decay element is the stable element into which a radioactive element breaks down. **The breakdown of a radioactive element into a decay element occurs at a constant rate.** Some radioactive elements decay in a few seconds. Some take thousands, millions, or even billions of years to decay. But no matter how long it takes for an element to decay, the rate of decay for that element is absolutely steady. No force known can either speed it up or slow it down.

Scientists measure the decay rate of a radioactive element by a unit called **half-life.** The half-life of an element is the time it takes for half of the radioactive element to decay.

For example, if you begin with 1 kilogram of a radioactive element, half of that kilogram will decay during one half-life. So at the end of one half-life, you will have 0.50 kilograms of the radioactive element and 0.50 kilograms of the decay element. Half of the remaining element (half of a half) will decay during another half-life. At this point one quarter of the radioactive element remains. How much of the decay element is there? This process continues until all the radioactive element has decayed. Figure 1–15 on page 26 illustrates the decay of a radioactive element with a half-life of 1 billion years.

If certain radioactive elements are present in a rock or fossil, scientists can find the absolute age of the rock or fossil. For example, suppose a rock contains a radioactive element that has a half-life of 1 million years. If tests show that the rock contains

Figure 1–14 *A knowledge of life science was needed to correctly reassemble the tiny bones in the skeleton of a hatching dinosaur. How does a knowledge of physical science help a scientist to determine the absolute age of the fossil?* ❷

F ■ 25

Radioactive dating of moon rocks brought back by the Apollo missions shows them to be 4.0 to 4.6 billion years old. The oldest rocks brought back from the moon are more than 0.5 billion years older than the oldest known Earth rocks. Scientists have evidence that the Earth and moon may have formed at about the same time. So from the age of rocks on the Earth and moon, scientists believe that the Earth is about 4.6 billion years old.

HISTORICAL NOTE
LEONARDO DA VINCI

Leonardo da Vinci, a fifteenth-century Italian artist and scientist, was the first person known to have expressed the opinion that fossils were deposited upon and pressed into muddy sediments.

1–2 (continued)

REINFORCEMENT/RETEACHING

Review the concept of half-life. Point out that half-life gives scientists the "clock" they need to measure back into geologic history. Describe the way half-life can be used to measure age through radioactive dating.

You may want to have students solve the following half-life problems.
- **A radioactive element has a half-life of 10 days. How much of an 8-g sample will be unchanged after 40 days?** (0.5 g.)
- **A fossil contains a radioactive element that has a half-life of 10,000 years. If the ratio of radioactive element to decay element is 1:3, how old is the fossil?** (20,000 years, or two half-lives.)

CONTENT DEVELOPMENT

Have students study the chart in Figure 1–15. Point out that 4 billion years ago there was 1 kg of the radioactive element. That is, the chart starts with a total of 1 kg.
- **How many kilograms of the decay element existed 4 billion years ago?** (0 kg.)

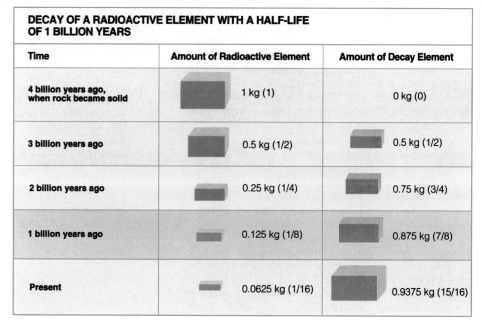

DECAY OF A RADIOACTIVE ELEMENT WITH A HALF-LIFE OF 1 BILLION YEARS

Time	Amount of Radioactive Element	Amount of Decay Element
4 billion years ago, when rock became solid	1 kg (1)	0 kg (0)
3 billion years ago	0.5 kg (1/2)	0.5 kg (1/2)
2 billion years ago	0.25 kg (1/4)	0.75 kg (3/4)
1 billion years ago	0.125 kg (1/8)	0.875 kg (7/8)
Present	0.0625 kg (1/16)	0.9375 kg (15/16)

Figure 1–15 *The rate of decay of a radioactive substance is measured by its half-life. How much of the radioactive element remains after 2 billion years?* ❶

equal amounts of the radioactive element and its decay element, the rock is about 1 million years old. Since the proportion of radioactive element to decay element is equal, the element has gone through only one half-life. Scientists use the proportion of radioactive element to decay element to determine how many half-lives have occurred. If the rock contains three times as much decay element as it does radioactive element, how many half-lives have occurred? How old is the rock? ❷

Many different radioactive elements are used to date rocks and fossils. Figure 1–16 lists some radioactive elements and their half-lives. One radioactive element used to date the remains of living things is carbon-14. Carbon-14 is present in all living things. It can be used to date fossils such as wood, bones, and shells that were formed within the last 50,000 years. It is difficult to measure the amount of

- **Three billion years ago, how much of the radioactive element was left?** (0.5 kg.)
- **How much of the decay element existed 3 billion years ago?** (0.5 kg.)

Explain that there will always be a total of 1 kg. The amount of the radioactive element remaining plus the amount of the decay element must equal 1 kg.
- **Some time between 2 and 3 billion years ago there must have been 0.3 kg of the radioactive element left. How much decay element would have existed then?** (0.7 kg.)

● ● ● ● **Integration** ● ● ● ●

Use the discussion of moon rocks to integrate concepts about the Earth's moon into your lesson.

CONTENT DEVELOPMENT

Have students look at the chart in Figure 1–16 and read the caption.
- **According to the chart, which elements are found in rocks but not in fossils?** (Rubidium and thorium.)
- **Which elements are found in fossils but not in rocks?** (Carbon.)

HALF-LIVES OF ELEMENTS USED TO FIND THE AGE OF ROCKS AND FOSSILS

Element	Half-life	Used to Find Age of
Rubidium-87	50.00 billion years	Very old rocks
Thorium-232	13.90 billion years	Very old rocks
Uranium-238	4.51 billion years	Old rocks and fossils in them
Potassium-40	1.30 billion years	Old rocks and fossils in them
Uranium-235	713 million years	Old rocks and fossils in them
Carbon-14	5770 years	Fossils (usually no older than about 50,000 years)

Figure 1–16 *Each radioactive substance has a different half-life. These substances are used to measure the age of different rocks and fossils. What is the half-life of potassium-40? Can you explain why carbon-14 is not used to determine the age of dinosaur fossils?* ③

carbon-14 in a rock or fossil more than 50,000 years old because almost all of the carbon-14 will have decayed into nitrogen. Nitrogen is the decay element of carbon-14. Using the information in Figure 1–16, which radioactive elements would you choose to date a fossil or rock found in the oldest rocks on Earth? ④

The Age of the Earth

Scientists use radioactive dating to help determine the age of rocks. By finding the age of rocks, they can estimate the age of the Earth. Scientists have found some rocks in South Africa that are more than 4 billion years old—the oldest rocks found on the Earth so far.

Radioactive dating of moon rocks brought back by the Apollo missions shows them to be 4 to 4.6 ① billion years old. The oldest moon rocks, then, are more than a half billion years older than the oldest known Earth rocks. However, because scientists have evidence that the Earth and the moon formed at the same time, they believe that the Earth is about 4.6 billion years old.

ANNOTATION KEY

Answers

① 0.25 kg. (Interpreting charts)

② 2 half-lives; 2 million years old. (Making calculations)

③ 1.30 billion years; Carbon-14 is difficult to measure in fossils more than 50,000 years old. (Making inferences)

④ Rubidium-87 or thorium-232. (Interpreting charts)

Integration

① Earth Science: Earth's Moon. See *Exploring the Universe,* Chapter 3.

HISTORICAL NOTE

HERODOTUS

The age of the Earth has been a topic of speculation and investigation for thousands of years. Even as far back as 450 BC, the Greek historian Herodotus observed that each annual flood of the Nile left a thin layer of mud over the valley. He recognized that the thick deposits at the delta must have been built up by these annual floodings over thousands of years. Based on these observations, he tried to estimate the age of the Earth.

- **Which of the elements listed in the chart are familiar to you?** (Students will probably recognize uranium, potassium, and carbon.)
- **What element on the chart would be best to determine the age of the oldest rocks?** (Rubidium.)
- **Which element would be best for dating recent fossils?** (Carbon.)

ENRICHMENT

Explain that the numbers following the elements in Figure 1–16 indicate a particular isotope of each element. Remind students that atoms of an element having the same atomic number but different atomic masses are called isotopes of that element. Atoms of the various isotopes of an element, therefore, have the same number of protons and electrons but different numbers of neutrons.

INDEPENDENT PRACTICE

▶ *Activity Book*

Students who need practice with the concepts of this section should be provided with the chapter activity Constructing a Model of Radioactive Decay.

GUIDED PRACTICE

▶ *Laboratory Manual*

Skills Development

Skills: Applying concepts, making observations, making inferences

At this point, you may want to have students complete the Chapter 1 Laboratory Investigation in the *Laboratory Manual* called Using the Rock Record to Interpret Geologic History. In the investigation students will observe and interpret subsurface rock layers.

1–3 A Trip Through Geologic Time

MULTICULTURAL OPPORTUNITY 1-3

To help your students to understand time, have them make a personal time line. Use a strip of adding machine paper cut to a length in centimeters that matches each student's age in years. (For example, a 12-year-old student would cut a 12-cm length of adding machine paper). Have students mark their strips with a line for each centimeter, representing each of their birthdays. Next have them add some significant dates to the time line—the first day of school, the birth of a younger brother or sister, a favorite vacation trip. Point out how this strip of adding machine paper is a record of each student's history, much like the geological time line is a record of Earth's history.

ESL STRATEGY 1-3

Suggest that students make a study chart of the major geologic eras and the changes on Earth attributed to them. They can use these headings: Geologic Era, Time Span, Period, Forms of Life.

Have students write an imaginary version of what happened to reduce the sea levels, dry up the rivers, flood the plains, and destroy the sea and land monsters that had dominated the Earth during most of the Mesozoic Era.

PROBLEM ??? Solving

People-Eating Dinosaurs?

As the world's leading expert on fossils, you have been called upon to resolve a growing controversy. Recently, a collection of human bones have been found at the mouth of an ancient river. Grooves on the bones show that they were chewed by a large animal. Near the bones were discovered the tracks of a meat-eating dinosaur. Newspapers throughout the world have declared the find as evidence that people and dinosaurs once lived together and that the dinosaurs hunted and ate people. However, as a scientist, you know that the dinosaurs were extinct for over 60 million years before the first humans evolved on Earth.

Interpreting Evidence

1. What tests would you perform to demonstrate that the humans were not eaten by dinosaurs?

2. What other hypothesis can you provide to explain the find?

1–2 Section Review

1. How is the law of superposition used to date fossils?
2. Do index fossils provide evidence for relative age or absolute age? Explain.
3. Compare an intrusion and an extrusion.
4. How do scientists use the half-life of a radioactive element to date rocks and fossils?

Connection—*You and Your World*
5. While digging in her backyard, Carmela finds the bones of a fish. Carmela immediately decides that the bones provide evidence that the area she lives in was once under water. Is Carmela correct in her analysis or could there be some other explanation for her findings?

1–2 (continued)

INDEPENDENT PRACTICE

Section Review 1-2

1. The law of superposition helps to determine whether a layer of rock is older or younger than another layer of rock. The ages of fossils can be compared by using the relative ages of the layers in which they are found.

2. Index fossils provide evidence for relative age. A layer of rock that includes an index fossil is close in age to another layer with the same index fossil. Or the layer might be older or younger than another layer.

3. An intrusion is hardened magma that has forced it way into layers of rocks. An extrusion is hardened magma that forms on the Earth's surface.

4. Scientists use the ratio of radioactive element to decay element to determine the number of half-lives that have passed since a rock or fossil was formed.

5. No, Carmela is not justified in her conclusion. There are a great many other possible explanations for the fish bones.

REINFORCEMENT/RETEACHING

Review students' responses to the Section Review questions. Reteach any material that is still unclear, based on their responses.

1–3 A Trip Through Geologic Time

The Earth's clock began ticking long ago. So long ago that the major units of time of this clock could not be seconds, hours, days, weeks, months, or even years. There would be just too many of these units for them to be useful in setting up a calendar of Earth's history. For example, more than 1.5 trillion days have passed since Earth formed. If each of these days took up one page of an ordinary office calendar, the calendar would be about 140,000 kilometers thick. Not very practical—and a bit difficult to carry around.

In order to divide geologic time into workable units, scientists have established the geologic time scale. **Earth's history on the geologic time scale is** ❶ **divided into four geologic eras: Precambrian Era, Paleozoic Era, Mesozoic Era, and Cenozoic Era.** An era is the largest division of the geologic time scale. Eras are broken into smaller subdivisions called periods.

The eras of Earth's geologic time scale are of different lengths. Geologic time is the length of time Earth has existed. The Precambrian (pree-KAM-bree-uhn) Era is the longest era. It lasted about 4 billion years and accounts for about 87 percent of Earth's history. The Paleozoic (pay-lee-oh-ZOH-ihk) Era

Guide for Reading

Focus on these questions as you read.

▶ *What are the four geologic eras?*
▶ *What were the major life forms and geologic events during the four geologic eras?*

Figure 1–17 *Many unusual animals lived long ago, such as the meat-eating saber-toothed cat (top) and the plant-eating dinosaur (bottom).*

F ■ 29

PROBLEM SOLVING
PEOPLE-EATING DINOSAURS?

In order to demonstrate that the humans were not eaten by dinosaurs, students should suggest tests to show that the grooves on the bones were not made by an animal of the dinosaur age. Dating processes might be used to explore this hypothesis. Casts of the nearby tracks might also be taken to determine if they are similar to any tracks proven to have been made by dinosaurs.

Another hypothesis that explains the find is that the groove marks on the human bones may have been made by an animal long after the human had died. Perhaps some animal of the modern age gnawed on the bones. The fact that the bones were found near the tracks may also have been a coincidence. The bones may have been carried there by a scavenger.

After students complete this Problem Solving feature, it may be useful to discuss the way the media interpret scientific material.

CONTENT DEVELOPMENT

Use the Focus/Motivation discussion to introduce the concept of a trip through geologic time. Define the term *geologic time* by explaining to students that geologic time is the long period of time the Earth has existed. Discuss the division of geologic time into eras and periods. Explain that there are also smaller subdivisions of geologic time: Periods can be divided into epochs; epochs can be divided into ages; and ages can be divided into phases.

● ● ● ● **Integration** ● ● ● ●

Use the discussion of the geologic time scale to integrate social studies concepts into your lesson.

CLOSURE

▶ *Review and Reinforcement Guide*
Have students complete Section 1–2 in the *Review and Reinforcement Guide.*

TEACHING STRATEGY 1–3

FOCUS/MOTIVATION

Begin by asking students the following.
• **Suppose you could go back in time. What time in the Earth's history would you most like to visit?** (Accept all answers.

Some students may wish to visit periods in recent history such as the Middle Ages; others may choose a time such as the age of dinosaurs; still others may wish to witness the formation of the Earth. List students' answers on the chalkboard. Then challenge students to write on a sheet of paper the times listed in chronological order. Students should use whatever previous knowledge they have, but they should not use their text. Ask students to put their lists aside. They will be referred to later in the section.)

Figure 1–18 describes the formation of many important mountain ranges. Have students locate each of these mountain ranges and draw them on a world map. Then have students use colors, labels, or some other technique to indicate the time in the Earth's history during which each mountain range was formed.

Various periods of the Paleozoic Era were named for locations, groups of people, cities, rivers, and mineral deposits. For example, the Cambrian and Silurian periods were named for Welsh tribes; the Carboniferous Period for coal beds; the Devonian Period for the English county of Devon; the Permian Period for the city of Perm in the Soviet Union; and the Mississippian Period for the Mississippi River valley.

1–3 (continued)

REINFORCEMENT/RETEACHING

Have students make up a simple chart showing the hierarchy of geologic time-unit divisions, which they often tend to confuse. The chart should show periods as subdivisions of the era and epochs as subdivisions of periods.

CONTENT DEVELOPMENT

Review this chart carefully with students, first explaining its overall structure and features, and then using it to review particular information on each of the four eras. Draw students' attention to the periods of the eras and to their particular characteristics. Do not attempt to hold students responsible for too much detailed information on tests, however. Instead, use the chart as a way of expanding your teaching and of providing the occasion for a broader learning experience for students.

GEOLOGIC HISTORY OF THE EARTH

Era	Precambrian	Paleozoic
Began (millions of years ago)	4600	570
Ended (millions of years ago)	570	225
Length (millions of years)	4030	345

Period	None	Cambrian	Ordovician	Silurian	Devonian
Began (millions of years ago)	4600	570	500	430	395
Ended (millions of years ago)	570	500	430	395	345
Length (millions of years)	4030	70	70	35	50

Earth's history begins; seas form; mountains begin to grow; oxygen builds up in atmosphere; first life forms in sea; as time passes, bacteria, algae, jellyfish, corals, and clams develop

Shallow seas cover parts of continents; many trilobites, brachiopods, sponges, and other sea-living invertebrates are present

Many volcanoes and mountains form; North America is flooded; first fish (jawless) appear; invertebrates flourish in the sea

Caledonian Mountains of Scandinavia rise; coral reefs form; first land plants, air-breathing animals, and jawed fish develop

Acadian Mountains of New York rise; erosion of mountains deposits much sediment in seas; first forests grow in swampy areas; first amphibians, sharks, and insects develop

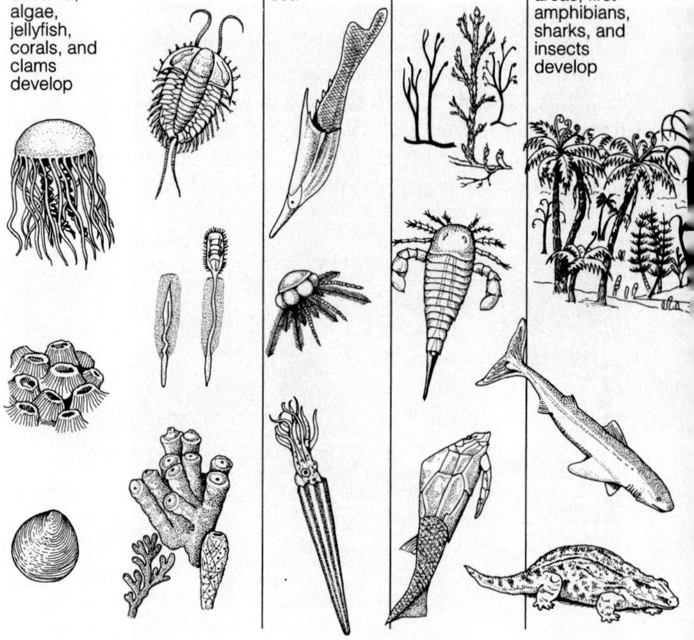

Figure 1–18 This chart illustrates the geologic history of the Earth. What events occurred during the Permian Period? When did modern humans appear? ❶

GUIDED PRACTICE

Skills Development

Skill: Interpreting charts

Have students observe Figure 1–18. Point out that unlike a time line, the chart is not drawn to scale horizontally.

• **How different would the chart look if it were drawn to scale?** (The Precambrian Era would take up the most space because it lasted much longer than the other eras.)

Direct students' attention to the time measurements that appear in the chart. Stress that the numbers given are in millions of years. In each era and period the top row of numbers tells how many millions of years ago each began, and the middle row tells how many millions of years ago the time division ended. The third row of numbers shows the lengths of the era or period.

• **How can you verify the length of each era or period?** (By subtracting the middle number from the top number.)

		Mesozoic			Cenozoic	
		225			65	
		65			The present	
		160			65	

*In North America, the Carboniferous Period is often subdivided into the Mississippian Period (345–310 million years ago) and the Pennsylvanian Period (310–280 million years ago).

Carboniferous*	Permian	Triassic	Jurassic	Cretaceous	Tertiary	Quaternary
345	280	225	190	136	65	1.8
280	225	190	136	65	1.8	The present
65	55	35	54	71	63.2	1.8
Appalachian Mountains of North America form; ice covers large areas of the Earth; swamps cover lowlands; first mosses, reptiles, and winged insects appear; great coal-forming forests form; seed-bearing ferns grow	Ural Mountains of Russia rise; first cone-bearing plants appear; ferns, fish, amphibians, and reptiles flourish; many sea-living invertebrates, including trilobites, die out	Palisades of New Jersey and Caucasus Mountains of Russia form; first dinosaurs and first mammals appear; modern corals, modern fish, and modern insect types develop	The Rocky Mountains rise; volcanoes of North American West are active; first birds appear; palms and cone-bearing trees flourish; largest dinosaurs thrive; primitive mammals develop	First flowering plants appear; placental mammals develop; dinosaurs die out, as do many sea-living reptiles	Andes, Alps, and Himalayan Mountains rise; first horses, primates, and humanlike creatures develop; flowering plants thrive; mammals take on present-day features	Ice covers large parts of North America and Europe; Great Lakes form as ice melts; first modern human beings appear; woolly mammoths die out; civilization begins

FACTS AND FIGURES

DATING ERRORS

Scientists estimate that radioactive dating of rocks 1 billion years old may be in error by only plus or minus 20 million years. Although this may sound like a large error, 20 million is only 2 percent of 1 billion, which is really a very small error.

based on the idea of faunal succession, which states that the fauna (animal life) of the past followed a specific order of succession.

CONTENT DEVELOPMENT

Have students study the chart on page 31.

• **In which era did the first winged insects appear? In which period?** (Paleozoic; Carboniferous.)

• **In which era did the first flowering plants appear? In which period?** (Mesozoic; Cretaceous.)

Now have students look over the entire two-page chart.

• **Which of the geologic eras is divided into the greatest number of periods?** (The Paleozoic Era is divided into six different periods.)

• **What would be the advantage of a chart that just showed four parts, one part for each era? What would be the disadvantages of such a chart?** (Accept all logical responses. A four-part chart would be simpler, but it could not include as much detail.)

INDEPENDENT PRACTICE

▶ *Activity Book*

Students who need practice with the concepts of this section should be provided with the chapter activity Life in the Geologic Time Scale.

CONTENT DEVELOPMENT

Have students look first at the part of the chart on page 30.

• **How many of the organisms can you identify in the pictures shown on page 30?** (Accept all logical responses.)

• **How many geologic eras are shown in this part of the chart? What are their names?** (Two eras are shown, the Precambrian and part of the Paleozoic.)

• **How many geological periods are shown in this part of the chart?** (Four periods are shown.)

• **In which era did the first fish appear? In which period?** (Paleozoic; Ordovician.)

• **In which period did the first forests grow? In which era?** (Devonian; Paleozoic.)

REINFORCEMENT/RETEACHING

Organic evidence used to divide geologic time is also used to determine whether geologic formations are of the same age. Such formations are assumed to be contemporary if they contain similar fossils. This assumption, in turn, is

SOIL FORMATION

Soil is to a large extent the creation of living things. Some organisms produce acids that chemically break down rocks. And certain bacteria cause the decay of dead animals and plants, forming material called humus. Larger animals such as moles, earthworms, ants, and beetles also help to break apart large pieces of soil and speed up the weathering process of the underlying rock.

Although scientists are not certain which plants were the first to live on land, it seems fairly clear that one of their "jobs" would have been helping to create the soil needed for more complicated plants.

1–3 (continued)

FOCUS/MOTIVATION

Show students pictures of various classes of organisms that originated during the Precambrian and the Paleozoic eras. These include jellyfish, worms, mollusks, fishes, and amphibians. Ask students to identify each and to describe its characteristics. Point out that the ancestors of the modern-day organisms had many characteristics in common with their descendants and also some characteristics that were unique to them.

CONTENT DEVELOPMENT

Briefly review the concepts of mountain building, volcanic acitivty, and continental drift as you discuss the changes occurring during the Precambrian and Paleozoic eras. Then go on to discuss life forms characteristic of these eras.

● ● ● ● **Integration** ● ● ● ●

Use the photograph of a volcanic eruption to integrate volcanoes into your lesson.

FOCUS/MOTIVATION

Suggest that students pick an era and write a short story about the adventures of a group of people suddenly transported to that era. Stories should be imaginative

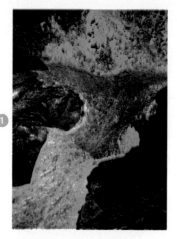

Figure 1–19 *Many volcanic eruptions occurred during the early geologic history of the Earth.*

Figure 1–20 *Modern sponges are quite similar to their earliest ancestors, which appeared at the end of the Precambrian Era.*

---

lasted about 345 million years, and the Mesozoic (mehs-oh-ZOH-ihk) Era about 160 million years. The Cenozoic (see-nuh-ZOH-ihk) Era, the era in which we now live, has lasted for only 65 million years. To get a better understanding of these eras and the events that occurred during each one, you will now take a short imaginary trip through Earth's history.

The Precambrian Era: The Dawn of Life

You begin your trip in space, gazing upon planet Earth from a safe distance. The time is 4.6 billion years ago, and Earth is a planet of molten rock much too hot to set foot upon. The atmosphere, unlike today's atmosphere, contains mainly poisonous gases. Because of widespread volcanic eruptions and cooling and hardening of lava, there is no record in the rocks from this distant time.

Skip ahead in your mind about 1.5 billion years later. The Earth is cooler now. Continents and oceans have formed. Although the air is warm and quite humid, little oxygen exists in the atmosphere. In fact, much of the atmosphere contains sulfur dioxide gas that has been released by volcanic eruptions. It is a dynamic and restless time in Earth's history.

Gazing down on the shores of a Precambrian sea, you notice a rocky shoreline with no signs of life. The scene is the same all over Earth. Rain falls and thunder rolls. But on land there are no plants to receive the rain and no animals to hear the thunder. Nor will there be for the next 2.6 billion years!

But there is life in the seas. If you look closely you will see faint signs of it. Lying on rocks just beneath the sea surface are patches of what looks like mold. The patches are several centimeters across. They are made of millions of bacteria clumped together in a tangled mat of threadlike fibers. Billions of years later scientists will find fossils of these bacteria—the oldest fossils ever found.

Plants related to modern seaweed are in the seas as well. They use simple chemicals in the water plus energy from sunlight to make their own food, just as green plants have done ever since. In the process, they produce oxygen. Over the next billion years, more and more oxygen will dissolve in the sea water

---

but also represent the characteristic organisms and geologic activity of the era students select.

CONTENT DEVELOPMENT

Explain that no fossils dating from the Precambrian Era were discovered until quite recently. Because of this, the Precambrian Earth was until fairly recently considered to be without life.

• **Why do you think Precambrian fossils were so difficult to find?** (Only invertebrates existed in Precambrian times, so

there were no hard, durable bony parts left by Precambrian organisms, which were small and soft-bodied. The extreme age of the Precambrian also made for deeper subsequent burial and also a greater time over which breakdown or destruction of fossils could occur.)

● ● ● ● **Integration** ● ● ● ●

Use the discussion of early life in the sea to integrate concepts of the origin of life into your lesson.

and enter the atmosphere. Animal life will become possible.

By the end of the Precambrian Era, animals such as jellyfish and worms have appeared in the seas, along with sponges and corals. Although sponges and corals look like plants, they are actually colonies of animal cells.

The Paleozoic Era: Life Comes Ashore

Your imaginary trip has brought you to the Paleozoic Era. The time is 570 million years ago, and even a quick glance alerts you to the fact that the land is still lifeless. But life is abundant in the seas. Worms of many kinds crawl across the sandy bottom. Strange "plants," which seem to grow from the sea floor, resemble animal horns, vases, and bells. These formations are actually sponges.

Parts of the sea floor contain lampshade-shaped shells. The shells have two parts that close to cover and protect the soft animal within. These animals are brachiopods.

Other sea animals have large heads, long thorny spines, and many body divisions. They have jointed legs like those of a modern insect or lobster. These animals are trilobites. They will become an important index fossil for the Paleozoic Era. They evolved rapidly during that era, leaving different forms in

Figure 1–21 *Brachiopods were quite common during the Paleozoic Era—about 30,000 species are known from their fossil shells. Although brachiopods look somewhat like clams, they are as distantly related to clams as you are.*

Figure 1–22 *Strange armored fishes swam in the waters during the Devonian Period. What is this period's nickname?* ❶

F ■ 33

INTEGRATION
LIFE SCIENCE

The Earth became suitable for animal life because of photosynthesis in plants. During photosynthesis, plants use carbon dioxide in the presence of sunlight to make food. A waste product of photosynthesis is oxygen. Animals need oxygen in order to carry out the process of respiration. As photosynthesis continued for millions of years, enough oxygen was released into the ocean and eventually into the atmosphere to sustain animal life. On the Earth today, the oxygen–carbon dioxide cycle is maintained by the taking in of carbon dioxide and the release of oxygen by plants, and the taking in of oxygen and the release of carbon dioxide by animals.

REINFORCEMENT/RETEACHING

One of the most important events in the history of life occurred during the Precambrian Era. This was the evolution of multicelled organisms.

Remind students that the first living organisms had only one cell. Although they appeared well over a billion years ago, it is not until 700 million years ago that multicelled animals are found in the fossil record.

Multicellularity has many advantages. The organism can have a larger size and a greater metabolic efficiency. Even more important, a multicelled animal can have specialized organs and make use of a sort of "division of labor."

Ask students to create a description of how life on Earth today would be different if multicelled organisms had not evolved.

GUIDED PRACTICE

Skills Development

Skill: Interpreting photographs

Have students observe Figure 1–21.
• **What organism is shown in this photograph?** (Students may realize that the brachiopods are invertebrates, or animals without backbones.)

Explain that *Brachiopoda* is a phylum of invertebrates that has existed since early geologic times. One member of this phylum is the lamp shell, a marine organism with two shells similar to some of the mollusks. The brachiopods are interesting to scientists because they have changed so little during many millions of years.

ENRICHMENT

▶ *Activity Book*

Students who have mastered the concepts in this section will be challenged by the chapter activity Interpreting an Ancient Puzzle.

Geologic changes occur slowly. Billions of years passed before the Earth's atmosphere contained enough oxygen to support animal life. The Earth's climate today, capable of sustaining an enormous diversity of plant and animal life, was achieved over a very long period of time.

Human activities can threaten the balance of the Earth's climate, perhaps changing it in ways that will endanger all organisms. And the life-threatening climatic changes could occur in an extremely short span of time.

Possible threats to the climate include the release of carbon dioxide, which could raise the temperature of the atmosphere, and the production of sulfur oxides and nitrogen oxides, which cause acid rain.

Figure 1–23 *Creatures of the Paleozoic seas included starfish (top) and trilobites (center and bottom). This starfish is unusual because it has six arms, whereas most modern starfish have five. Why are trilobites important index fossils?* ❶

each of the era's periods, and then became extinct by the era's end.

The seas are truly teeming with life. Fishes can be found almost everywhere. Fishes are the first vertebrates, or animals with backbones, to appear on Earth. In fact, the Devonian period, one of the periods of the Paleozoic Era, is often called the Age of Fishes.

By the end of the Paleozoic Era, the land is no longer lifeless. Huge forests of ferns have developed. There are also cycads, which are trees with a crown of fernlike leaves. Cycads are among the first seed plants. In the future, the sago palm, one of the few modern cycads, will be called a living fossil. Scientists believe that the remains of these forests of ferns and other plants formed the huge coal deposits in the United States and other parts of the world.

Amphibians, such as *Eryops*, now appear as well. Amphibians are the first land vertebrates. The name amphibian, which means "living a double life," is quite appropriate, since amphibians typically spend their early lives in water and then move to land. *Eryops* is a far larger amphibian than its twentieth-century relatives, such as frogs and toads. It is more than 2 meters long, with a large head and a thick, clumsy body. *Eryops* waddles through the forest in search of king-sized roaches and other tasty meals. But because *Eryops* must keep its skin moist in order to survive, it does not move far from water.

By the end of the Paleozoic Era, the amphibians have run up against hard times. There is drought, and the climate has cooled. Mountains are rising in what will become Norway, Scotland, Greenland, and parts of North America. These areas, as well as all the other landmasses on Earth, are joined together as one single continent. Scientists will one day call ❶ this continent Pangaea, but that is still about 225 million years in the future.

New kinds of animals that live on land all the time are appearing. Their tough skin is protected by scales or hard plates. Unlike amphibians, these animals do not lose water through their skins. Their eggs have thick shells, so the eggs do not dry out. The animals are reptiles, and for the next 160 million years they will dominate Earth.

1–3 (continued)

CONTENT DEVELOPMENT

Explain that the first land vertebrates appeared during the Paleozoic Era. These were the amphibians. Have students observe Figure 1–24 and read the caption.
• **What organism is shown in the illustration?** (An amphibian called *Eryops.*)
• **How does this animal differ from modern frogs and toads?** (It is much bigger, more than 2 meters long.)

Explain that most amphibians spend the early stages of their lives in water, breathing through gills. Then the animal develops into an adult form with lungs. Adult amphibians are still able to stay under water for long periods because their skin also functions as a respiratory organ. Other examples of amphibians include salamanders, mud puppies, and blindworms.

REINFORCEMENT/RETEACHING

Some students may have difficulty appreciating the vast periods of time being discussed in this section. To help them understand this concept, have them imagine that the age of the Earth were given in common time units:

4.6 billion years
= 55.2 billion months
= 239.2 billion weeks
= 1674.4 billion days
= 40,185.6 billion hours
= 2,411,136 billion minutes
= 144,668,160 billion seconds

CONTENT DEVELOPMENT

Explain that major changes occurred in the Earth near the end of the Paleozoic Era. Mountains arose and various landmasses joined together into one continent.

New kinds of animals also appeared, and one of these is the reptile. Explain that reptiles arose from primitive amphibians. Unlike their ancestors, reptiles

Figure 1–24 *Amphibians evolved from fish ancestors during the Paleozoic Era. Like typical amphibians, Eryops lived in water when it was young. As an adult, this ancient amphibian lived in moist places on land.*

The Mesozoic Era: Mammals Develop

Your trip continues as you enter the Mesozoic Era, which began about 225 million years ago. The Mesozoic Era is a period of many changes—both in the land and in the living things that inhabit Earth. Scientists believe that Pangaea began to break apart during the Mesozoic Era. The expansion of the ocean floor along midocean ridges caused the continents gradually to spread apart. Midocean ridges are chains of underwater volcanic mountains. Today the continents are still moving apart at the midocean ridges.

As Pangaea broke apart, there were numerous earthquakes and volcanic eruptions. Many mountains were formed at this time. The Appalachian Mountains were leveled by erosion during the early part of the Mesozoic Era. Then they were uplifted again late in the era. The Sierra Nevada and Rocky Mountains were formed during the late stages of the Mesozoic Era.

Scientists divide the Mesozoic Era into three periods. The oldest period is called the Triassic (trigh-AS-ihk) Period. The middle period is called the Jurassic (joo-RAS-ihk) Period. The youngest period is called the Cretaceous (krih-TAY-shuhs) Period.

ACTIVITY READING

Good Mother Lizard

If you think that dinosaurs were slow-witted, slow moving creatures, you will be in for a surprise if you read *Maia—A Dinosaur Grows Up* by Jack Horner. In this book you will discover what it might have been like to be an infant dinosaur in a world with so many ferocious creatures.

F ■ 35

ACTIVITY READING

GOOD MOTHER LIZARD

Skill: Reading comprehension

For many years scientists believed that dinosaurs were somewhat indifferent parents. These beliefs were based on the fact that few modern-day reptiles pay much attention to their offspring.

Dinosaur remains found in Montana in the late 1970s have led paleontologists to change their view. They now believe that at least some dinosaurs actively nurtured their young, bringing them food in a birdlike fashion.

After students have read Horner's book, lead a class discussion on why it is so difficult to be certain of the habits and behaviors of the dinosaurs. What problems would this have posed for Horner while he was writing his book?

Integration: Use this Activity to integrate language arts into your lesson.

could break away from an aquatic existence because they had shelled eggs and internal fertilization.

● ● ● ● **Integration** ● ● ● ●

Use the discussion of continental drift to integrate Pangaea into your lesson.

ENRICHMENT

Have students use reference materials to list as many modern-day reptiles as they can. Possible answers include crocodiles, alligators, turtles, lizards, and snakes.

CONTENT DEVELOPMENT

Describe the general characteristics of the Earth's surface and its major organisms during the Mesozoic Era. Point out that although large reptiles were the dominant life forms, there were many small mammals and vegetation that were quite similar to our own, especially during the Cretaceous Period. If possible, provide illustrations, in the form of artists' rendering, of some of these life forms.

● ● ● ● **Integration** ● ● ● ●

Use the discussion of the expansion of the ocean floor to integrate concepts of oceanography into your lesson.

ENRICHMENT

Have students do basic library research on one or more of the organisms that lived during the Mesozoic Era. They can prepare a report on the subject and present it to the class.

ACTIVITY
DOING
PANGAEA

Skills: Applying concepts, making inferences, making observations

Material: world map

This activity helps students visualize the landmass of the Earth as it existed many millions of years ago and provides them with more information on fossil history.

Students will find that scientists believe Pangaea split into two large continents called Gondwanaland and Laurasia. Gondwanaland then split into three parts, one with South America and Africa, a second with Antarctica and Australia, and a third consisting of India. North America and Eurasia were formed when Laurasia split apart.

Have students draw maps of Pangaea and display them to the class while reporting on their findings.

1–3 (continued)

FOCUS/MOTIVATION

Because most students are fascinated by dinosaurs, have them construct shoebox dioramas showing the characteristics of the Mesozoic Era.

CONTENT DEVELOPMENT

The largest of the meat-eating dinosaurs, *Tyrannosaurus rex,* is probably the most familiar to students. They may have seen a skeleton or life-sized model of this dinosaur in a museum. It is interesting to note that scientists have never found a complete skeleton for this creature, so they have only a reasonable guess as to what it looked like.

ENRICHMENT

Scientists have found fossil evidence of flying reptiles that lived during the Mesozoic Era. Have interested students find out what some of these animals looked like and prepare drawings to share with the class.

Figure 1–25 *Reptiles dominated the land, sky, and seas during the Mesozoic Era. Porpoiselike ichthyosaurs and long-necked plesiosaurs swam swiftly through the oceans in search of food.*

ACTIVITY
DOING

Pangaea

Fossils of the same land animals have been found on separate continents. Scientists say such findings are evidence that at one time Earth had one supercontinent called Pangaea.

Using materials in the library, find out which fossil animals support the idea of Pangaea. Write down the names of the fossil animals and where they were discovered on the continents.

Mark off these discovery sites on a world map. At which points do you think the continents came together?

36 ■ F

THE TRIASSIC PERIOD As you enter the Triassic Period you notice that the drought that began in the Paleozoic Era has not ended. The climate, in fact, is even hotter than before. Slowly, very slowly, you see North and South America begin to separate from Africa. A narrow sea opens between North America and what is now Iceland and England. This sea will become the North Atlantic Ocean. The southern lands of Africa, South America, Antarctica, and India are still joined.

In the seas, you spot creatures that are shaped like fish. But they are not fish. The bones in their fins have five fingerlike projections, like the limbs of land animals. They have lungs and breathe air. These creatures are reptiles—land-living animals—that have returned to the sea.

Mammals appear in the Triassic Period. Mammals are animals with hair or fur, whose offspring, for the most part, do not hatch from eggs. The young grow and mature in their mother's body before birth.

Ferns and seed ferns are still common. Cycads are growing bigger. During this time, the trees that will become the Petrified Forest of Arizona are

CONTENT DEVELOPMENT

Discuss possible reasons for the extinction of the dinosaurs. Ask students what theory they favor and why. In doing so, encourage them to articulate their reasons clearly and to support them on the basis of logical, scientific reasoning.

• **Why do you think other life forms, such as mammals and flowering plants and trees, were able to survive the changing conditions that brought about the extinction of the dinosaurs?** (Answers will vary considerably, as do proposed answers

to these questions within the scientific community. The small size, warmbloodedness, and perhaps greater intelligence and adaptability of mammals may have accounted in part for their survival.)

Direct students' attention to Figure 1–25, which illustrates one of the strange organisms that lived in the past.

• **What other examples of extinct organisms can you name and describe?** (Answers may include mammoths, various dinosaurs, and various sea creatures and primitive plants.)

uprooted by floods. The first dinosaurs are appearing. Many are small, no bigger than chickens. They have small heads, long tails, and walk on their hind legs.

THE JURASSIC PERIOD As you move forward into the Jurassic Period, you notice that the dinosaurs that will be found at Bone Cabin Quarry have now appeared. The Age of Dinosaurs has begun. Volcanoes are active in the American West. The mountains of the Sierra Nevada and Rocky Mountain ranges are rising. The North Atlantic is still quite narrow. The southern continents, still closely linked, are just beginning to separate. Animals and plants are similar throughout this "supercontinent."

Huge cycads and modern-looking evergreens called conifers make up the forests. There are no flowers yet. Toward the end of the Jurassic Period, one of the first birds, Archaeopteryx (ahr-kee-AHP-ter-iks), appears. Its name means "ancient wing."

THE CRETACEOUS PERIOD The Cretaceous Period is a time of widespread flooding of continents by seas. The continents continue moving apart. By the end of the period, they are pretty much as they are now, although North America and Europe are still joined.

Dinosaurs still dominate the world. *Tyrannosaurus,* the greatest meat eater of all times, stalks the land. Among the plant eaters are the armored *Triceratops* and the strange-looking duck-billed dinosaurs. By the end of the Cretacious Period, however, all the dinosaurs will have died out. So will the sea-living reptiles. Of all the different reptiles of the Mesozoic Era, only crocodiles, turtles, lizards, and snakes will have survived. The mysterious mass extinctions of so many forms of life will puzzle scientists of the future. The scientists will debate whether the extinctions were due to a change in climate, a worldwide disease, or even the result of a gigantic asteroid crashing into the Earth.

The Cretaceous Period was a time of rapid change—rapid, that is, in terms of Earth's long past. During this period, sea levels dropped. Rivers and flood plains, where many dinosaurs thrived, dried up. Flowering plants appeared—among them such familiar trees as magnolia, oak, fig, poplar, elm,

---

CAREERS

Museum Technician

People who prepare museum collections are called **museum technicians.** They clean and preserve animal and plant specimens and carefully mount and arrange them in glass cases. They also work with fossils that have been recently discovered, restoring the skeletal parts with the use of clay, plaster, and other materials.

Museum technicians require special skills at doing detailed work, so on-the-job training is important. For further career information, write to American Association of Museums, 1055 Thomas Jefferson St., NW, Washington, DC 20007.

F ■ 37

ANNOTATION KEY

Integration
❶ Earth Science: Mass Extinction. See *Exploring Earth's Weather,* Chapter 2.

INTEGRATION
LANGUAGE ARTS

The three periods of the Mesozoic Era derived their names from quite different sources. The Triassic Period was named after threefold divisions of rock sequences. The Jurassic Period was named after the Jura Mountains in Europe. The Cretaceous Period took its name from *creta,* the Latin word for "chalk."

BACKGROUND INFORMATION
FLYING DINOSAURS

Scientists have found fossil evidence of flying animals that lived during the Mesozoic Era. *Pterodactyls* were flying reptiles. They had wings made of thin pieces of skin attached to their front legs and body, and they had a long, thin beak filled with sharp, curved teeth.

Another winged animal was *Archaeopteryx,* one of the first birds. This creature had wings with feathers but still had the skeleton and toothed beak of a reptile.

● ● ● ● **Integration** ● ● ● ●

Use the discussion of dinosaurs to integrate concepts of mass extinction into your lesson.

GUIDED PRACTICE

Skills Development

Skill: Making models

Have students attempt to create illustrations, based on drawings in books or magazines, of various dinosaurs. Alternatively, they can use modeling clay to make models of dinosaurs and can prepare a display of these creatures that includes vegetation and small "hills" or "volcanoes" (made of clay or sand) and that illustrates the dinosaurs in their natural habitats.

INDEPENDENT PRACTICE

▶ *Activity Book*

Students who need practice with the concepts of this section should be provided with the chapter activity Dinosaur Adaptations.

ENRICHMENT

Invite a local geologist or "rockhound" to visit your classroom to talk about methods for identifying rocks of different geologic ages in the field. Have him or her share information on where to look locally for rocks of different eras or periods.

Figure 1–26 *Dinosaurs were still masters of the land during the Cretaceous Period. But their reign was soon to come to an end.*

birch, and willow. As the new plant life spread and flourished, most of the great tree ferns and cycads died out. The world was beginning to take on a look that is much more familiar to you.

The Cenozoic Era: A World With People

You are about to enter the era that began approximately 65 million years ago—the era in which you live. The Cenozoic Era will be divided into two great periods known for the evolution of the first horses and the evolution of the first animals to walk on two feet. Great sheets of ice will sweep across the land. And finally, almost at the end of your trip through geologic time, a new kind of living thing will make its home on Earth and attempt to make sense of all that has passed.

THE TERTIARY PERIOD   You find yourself near what will one day be the town of Green River, in southwest Wyoming, not too many kilometers from the site of Bone Cabin Quarry. It is about 50 million years ago. Although in many ways the land is familiar, there is something odd about it.

In a grassy meadow, you hear birds singing. There are groups of redwoods, oaks, and cedars. But there are groves of palm trees, too! Worldwide, the

38 ■ F

1–3 (continued)

CONTENT DEVELOPMENT

Discuss the Cenozoic Era and its two periods, the Tertiary and the Quaternary. Contrast this era with the preceding ones as to dominant life forms.

Explain that this era is the one in which we currently live. The Tertiary Period is marked by the evolution of the first horses; the Quaternary period is marked by the evolution of the first animal to walk on two feet.

• **In which period do you think the first ancestors of humans appeared?** (The Quaternary.)

● ● ● ● **Integration** ● ● ● ●

Use the discussion of *Uintatherium* to integrate concepts of language arts into your lesson.

GUIDED PRACTICE

Skills Development

Skill: Making predictions

Ask students to consider what types of changes might signal an end to our present era, the Cenozoic. Also ask them to consider what factors might bring about these changes. For example, could environmental pollution eventually create catastrophic natural changes and bring our era to a close?

REINFORCEMENT/RETEACHING

Have students work in groups. Ask each group to find photographs or drawings of familiar living things such as a rose, a horse, coral, a bird, a jellyfish, a clam, an insect, and a human being.

Have students display their photographs or drawings in chronological order, showing when each organism first appeared on the Earth. (For example, clams first appeared during the Precambrian Era.)

climate is mild. And it will stay that way through most of this period.

Trotting through the meadow you see a beast that makes you wonder whether you are in Africa. It is about the size of an elephant, with elephantlike legs. Its skin is gray and wrinkled. Its tail, with a tuft of hair at the end, looks much like a lion's tail. But its head and ears are those of a rhinoceros. Instead of one or two horns, however, the animal has six horns.

The animal is *Uintatherium* (yoo-wihn-tuh-THEER-ee-uhm). The name means "Uinta beast," after the ① Uinta Mountains of Utah and Wyoming. Wandering the meadows with *Uintatherium* is a direct ancestor of the rhinoceros. It is no larger than a large dog. Another dog-sized animal, *Eohippus,* is the earliest horse.

Somewhere around 3.4 million years ago, toward the end of the Tertiary Period, humanlike creatures begin walking upright on the African plains. One is a small adult female about 1 meter tall. Scientists will find her skeleton in 1977. They will call her "Lucy."

THE QUATERNARY PERIOD The climate turns sharply colder during the Quaternary Period. On four different occasions, great sheets of ice advance from the Arctic and Antarctic regions, only to retreat

ACTIVITY
DOING

The Time of Your Life

1. On white index cards, write several important events that have happened to you in your life. *Place one event on each card.*

2. Arrange the cards in the order in which the events happened.

3. Using colored index cards, write one of the following on each: Preschool Years, Early Elementary School Years, Middle Elementary School Years, Late Elementary School Years, Junior High School/Middle School Years (if applicable). Insert each colored index card in front of the group of events that occurred during those years.

How does the arrangement of the cards resemble a geologic time line?

BACKGROUND INFORMATION
"LUCY"

The nearly complete fossil skeleton nicknamed "Lucy" represents an early hominid that anthropologists have placed in the genus *Australopithecus.* Current studies suggest that there were at least four different species in this genus. All of them lived between 4 and 1.4 million years ago, walked upright, and had much smaller brains than present-day humans. Many questions about these first primates remain to be answered.

ENRICHMENT

Relate the following situation to students.

A nineteenth-century oil shale mine in the center of one of West Germany's most populated areas has created a scientific controversy. The West German government proposes to eliminate this eyesore and solve its solid-waste disposal problem by filling the pit with garbage. Paleontologists, on the other hand, want the mine set aside for scientific research because it contains the fossilized remains of an entire ecosystem from the Eocene Epoch—when mammals were first established.

• **Should the government be allowed to carry out its plan for filling the fossil-rich pit with garbage? Explain your answer.** (Accept all reasonable answers. You may want to remind students that 99 percent of all species that ever lived are now extinct and that fossils provide the only means by which to study these organisms.)

• **What other solutions to this problem can you suggest?** (Accept all reasonable answers. Answers might include a compromise—scientific research for a certain period of time, then the pit could be filled with garbage; or scientific assistance in finding an alternative means of garbage disposal.)

ENRICHMENT

Have students choose one of the mammals, such as the horse, that appeared during the Cenozoic Era. Then ask students to research the various stages in this animal's development up to the present day. Students may wish to make sketches or other models of the various forms of the animal.

REINFORCEMENT/RETEACHING

Have students do library research on the changes that took place in their own region of the country during the Mesozoic and Cenozoic eras. These may include changes in land elevation, formation of local mountains or other land forms, and appearance and extinction of various organisms, as supported by the fossil record unearthed in their own region.

Students may wish to prepare simple maps, charts, or drawings to be displayed while they give an oral presentation or to be included with a written report.

Integration
1 Earth Science: Ice Age. See *Exploring Earth's Weather*, Chapter 2.
2 Earth Science: Continental Drift. See *Dynamic Earth*, Chapter 3.

Figure 1–27 Uintatherium, *which lived during the beginning of the Tertiary Period, was one of the largest and strangest-looking mammals ever to walk the Earth.*

1–3 (continued)

CONTENT DEVELOPMENT

Have students examine Figure 1–27 and read the caption.
• **How would you describe the animal shown in the illustration?** (Accept all logical answers.)
• **Do you think this animal might be the ancestor of any animal living today?** (Accept all logical answers.)

Media and Technology

At this point, students' understanding of the geologic events that have shaped Earth's structure and evolution will be enhanced if they begin their exploration of geologic time using the Interactive Videodisc called TerraVision. Using the Videodisc, students can help solve a geologic history puzzle that, upon completion, reenacts the sequence of geologic events over 500 million years.

Media and Technology

Students can explore the topic of continental drift and plate tectonics through simulation experiences by using the Interactive Videodisc/CD ROM entitled Science Discovery. Simulations and explanatory video clips of plate tectonics, continental drift, and plate boundary collisions will enable students to relate motions of Earth to geologic events that shaped the process of evolution.

● ● ● ● **Integration** ● ● ● ●

Use the discussion of the Quaternary Period to integrate the concept of ice age into your lesson.

REINFORCEMENT/RETEACHING

Have students work in pairs. Ask each pair to choose one period from the Pale-

Figure 1–28 *The weather became much cooler during the Quaternary Period. Sheets of ice repeatedly advanced and retreated over the Earth.*

as the climate becomes milder. At their worst, these ice ages are like winters that last thousands of years. 1 Large parts of Europe and North and South America are ice-covered. The last ice age ends about 11,000 years ago. As the world warms, farming becomes widespread and modern civilization begins.

Now it is time to end your trip and return home. What will happen in the Quaternary Period from this point on is modern history. The only way to find out is to wait and see.

1–3 Section Review

1. What are the four eras of geologic time?
2. In what era did the first vertebrates appear?
3. During what period did the first birds evolve?
4. Identify and describe one animal associated with each of the following eras: Paleozoic, Mesozoic, Cenozoic.
5. Why is it difficult to determine the geologic era to which an area of metamorphic rock belonged?

Critical Thinking—*Making Inferences*
6. How might the collision of a huge asteroid with Earth result in the extinction of the dinosaurs? *Hint:* Assume the dinosaurs were not killed by the impact alone.

ozoic, Mesozoic, or Cenozoic eras to research in detail. Students may wish to present their information in the form of a chart or time line.

GUIDED PRACTICE

Skills Development
Skill: Analyzing data

Have students conduct surveys to determine what a number of people believe the Earth's age to be, and what evidence the people have for their beliefs. Students

can construct simple bar-type histogram graphs to display numerical data on the numbers of people who (1) have no opinion, (2) believe the Earth to be more than 4 billion years old, (3) believe the Earth to be less than 10,000 years old, etc. Students can also prepare a report on the reasoning that was the basis for the various opinions.

ENRICHMENT

Have interested advanced students do library research on humanlike fossils and

CONNECTIONS

Breaking Up Is Hard to Do ❷

While we tend to think of the continents as being somewhat timeless, you have read that the continents are much different today than they were in the past. The idea that continents are in motion, moving along on huge plates that make up Earth's crust, is called *continental drift*.

Scientists believe that some 250 million years ago, all the world's landmasses were contained in one supercontinent called Pangaea. But about 200 million years ago, Pangaea split into two large continents called Gondwanaland and Laurasia. As time passed, Gondwanaland split into three parts. One part consisted of South America and Africa. Another part consisted of Antarctica and Australia. The third part was India. More time passed, and India drifted north and collided with Asia. South America and Africa separated. Laurasia, the other continent, split apart to form North America and Eurasia. Australia broke away from Antarctica and slowly drifted to its current position.

The continents are still drifting today at a rate of about 1 to 5 centimeters per year. What might the Earth look like in another 100 million years?

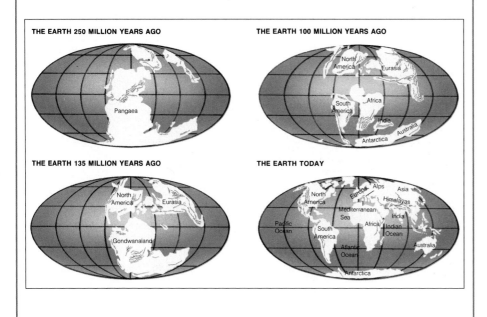

THE EARTH 250 MILLION YEARS AGO — Pangaea

THE EARTH 100 MILLION YEARS AGO — North America, Eurasia, Africa, South America, India, Antarctica, Australia

THE EARTH 135 MILLION YEARS AGO — North America, Eurasia, Gondwanaland

THE EARTH TODAY — North America, Europe, Alps, Asia, Himalayas, Mediterranean Sea, Africa, India, Pacific Ocean, South America, Indian Ocean, Atlantic Ocean, Australia, Antarctica

CONNECTIONS
BREAKING UP IS HARD TO DO

Before students read the Connections feature, it may be helpful to provide some historical information on theories about continental drift.

In the early 1900s a German scientist named Alfred Wegener became interested in the way the outlines of the continents seemed to fit together. He studied this peculiarity and eventually proposed that at one time the Earth had one giant landmass, called Pangaea, that split apart to form today's continents.

Today, Wegener's theory has been improved, and new evidence, such as the distribution of rocks on the continents and the theory of plate tectonics, continues to support his basic idea.

Students may have varying theories about what the Earth may look like 100 million years from now. They should incorporate the idea that the continents will continue to drift from their current positions to form different landmasses.

If you are teaching thematically, you may want to use the Connections feature to reinforce the themes of systems and interactions or patterns of change.

Integration: Use the Connections feature to integrate the theory of continental drift into your lesson.

6. Accept all logical responses. A very large asteroid would throw a great deal of dust and debris into the atmosphere. This would lower the temperature and change the climate. If the asteroid fell into an ocean, it could cause tidal waves or a rise in the sea level.

REINFORCEMENT/RETEACHING

Review students' responses to the Section Review questions. Reteach any material that is still unclear, based on their responses.

CLOSURE

▶ *Review and Reinforcement Guide*

Have students complete Section 1–3 in the *Review and Reinforcement Guide*.

on various theories that have attempted to describe human evolution and the possible interactions between different species or subspecies of humanlike organisms.

INDEPENDENT PRACTICE

Section Review 1–3

1. Precambrian Era, Paleozoic Era, Mesozoic Era, and Cenozoic Era.

2. The first vertebrates appeared in the Paleozoic Era.

3. The Jurassic Period.

4. Answers should be consistent with the information presented in the text.

5. Due to the intense heat and pressure involved in the formation of metamorphic rock, few or no fossils will be found to help date the geologic era in which the rock formed. Also, these same changes make it difficult to use radioactive dating techniques on metamorphic rocks.

Laboratory Investigation

INTERPRETING FOSSIL MOLDS AND CASTS

BEFORE THE LAB

1. Gather all materials at least one day prior to the investigation. You should have enough supplies to meet your class needs, assuming six students per group.
2. Be certain that you are familiar with the directions for mixing the plaster of Paris. Carry out a trial run, noting the amount of time required for the hardening of the plaster mixture.

PRE-LAB DISCUSSION

Before beginning this investigation, review with students the nature of sedimentation and of fossil mold and cast formation. Make sure that it is clear which of the steps in the investigation produces a mold and which produces a cast. Explain the mixing procedure for the plaster and give students some idea of the hardening time.

Laboratory Investigation

Interpreting Fossil Molds and Casts

Problem

What fossil evidence can be obtained from molds and casts?

Materials (per group)

```
small, empty milk carton
petroleum jelly
plaster of Paris
stirring rod or spoon
3 small objects
```

Procedure

1. Open completely the top of the empty milk container. Grease the inside of the container with petroleum jelly.
2. Mix the plaster of Paris, following the directions on the package. Pour the mixture into the milk container so that the container is half full.
3. Rub a coat of petroleum jelly over the objects you are going to use.
4. When the plaster of Paris begins to harden, gently press the objects into the plaster so that they are not entirely covered. After the mixture has hardened, carefully remove the objects. You should be able to see the imprints of your objects.
5. Coat the entire surface of the hardened plaster of Paris with petroleum jelly.
6. Mix more plaster of Paris. Pour it on top of the hardened plaster of Paris so that it fills the container. After the plaster hardens, tear the milk carton away from the plaster block. Gently pull the two layers of plaster apart. You now have a cast and a mold of the objects.

7. Exchange your molds and casts for those of another group. Number each cast and mold set from 1 to 3. In a chart, record the number of each set. Record your prediction of what object made each cast and mold.
8. Get the original objects from the other group and see if your predictions were correct. Record in your chart what the actual object is.

Observations

1. What are the similarities and differences between the casts and molds?
2. What are the similarities and differences between the casts and the original objects they were made from?

Number	Predicted Object	Actual Object

Analysis and Conclusions

1. Compare the formation of a plaster mold with the formation of a fossil mold.
2. Compare the way you predicted what the unknown object was with the way a scientist predicts what object left a fossil cast or mold.
3. **On Your Own** Find a set of prints or tracks left by an animal in concrete (such as a concrete sidewalk). Determine what the animal looked like, based on its "fossil" prints.

TEACHING STRATEGY

1. Have the teams follow the directions carefully as they work in the laboratory.
2. You may want to circulate through the room assisting students who have trouble mixing their plaster of Paris.

DISCOVERY STRATEGIES

Discuss how the investigation relates to the chapter ideas by asking open questions similar to the following.
• **Which types of objects do you think will make the best molds? The best casts?** (Accept all logical responses—making predictions.)
• **What types of information about an organism cannot be determined from molds, casts, and other fossils?** (Possible answers include internal organs, colors, and behavior—making inferences.)

OBSERVATIONS

1. Students should notice that a mold is an imprint of the outside of the object, whereas casts are the fillings of the mold after the object has been removed.
2. Students should notice some distortion in the shape of the cast as opposed to the actual object.

Summarizing Key Concepts

1-1 Fossils—Clues to the Past

▲ Fossils are the remains of once-living things.

▲ Fossils can form by the process of petrification, in which plant and animal parts are changed into stone.

▲ Fossils—in the form of molds, casts, and imprints—record the shapes of living things that have been buried in sediments.

▲ Fossils of entire animals are formed as the animals are buried in tar, amber, or ice.

▲ Trace fossils are any marks formed by an animal and preserved as fossils.

▲ Fossils indicate that many different life forms have existed throughout Earth's history.

1-2 A History in Rocks and Fossils

▲ The law of superposition states that in a series of sedimentary rock layers, younger rocks normally lie on top of older rocks.

▲ Scientists can tell the order in which past events occurred and the relative times of occurrence by studying sedimentary rock layers, index fossils, and unconformities.

▲ Index fossils are used to identify the age of the sedimentary rock layers containing them.

▲ The half-life of a radioactive element is the amount of time it takes for half the atoms in a sample of that element to decay.

▲ Scientists can determine the absolute age of rocks and fossils by using radioactive-dating techniques.

1-3 A Trip Through Geologic Time

▲ Scientists have set up a geologic calendar, divided into four eras: the Precambrian, the Paleozoic, the Mesozoic, and the Cenozoic.

▲ The Precambrian Era began 4.6 billion years ago. During this era, the first plant life formed in the seas.

▲ The Paleozoic Era began 570 million years ago. Sea animals, land plants, and land animals appeared during the Paleozoic Era. Reptiles were the first land animals able to survive out of water.

▲ The Mesozoic Era—containing the Triassic, Jurassic, and Cretaceous periods—began about 225 million years ago. During this time mammals, dinosaurs, and birds evolved.

▲ The Cenozoic Era, which includes the Tertiary and Quaternary periods, began 65 million years ago. During that time the first horses and the first humans evolved.

Reviewing Key Terms

Define each term in a complete sentence.

1-1 Fossils—Clues to the Past
fossil
sediment
petrification
mold
cast
imprint

trace fossil
evolve

1-2 A History in Rocks and Fossils
law of superposition
index fossil
unconformity

fault
intrusion
extrusion
half-life

F ■ 43

Part 1

You may wish to have some students prepare imprint fossils by pressing thin objects into the wet plaster surface and then removing the objects. Students can then also write a short report contrasting imprint fossils, molds, and casts.

Part 2

• **How would the molds and casts be affected by extreme conditions, such as very high temperatures or pressures? What does this suggest about fossil preservation in metamorphic rock?** (The molds and casts would probably be destroyed or seriously damaged. Metamorphic rocks, which have been subjected to such conditions, are not likely to contain well-preserved fossils.)

• **What would have happened if the "organism" had been placed into a medium at very high temperature, such as magma? What does this suggest about fossil preservation in igneous rock?** (The "organisms" would probably be completely destroyed. Igneous rock does not typically contain fossils.)

ANALYSIS AND CONCLUSIONS

1. The formation of the students' molds is similar to the formation of a fossil mold in that they are both made by filling in a cavity. The difference lies in the time it takes each mold to form. Fossil molds form after millions of years, whereas plaster molds form within a few hours.

2. Students should infer that their method is, in fact, quite similar to the methods scientists use when studying real molds and casts.

3. Accept all logical responses. Student answers will vary, depending on tracks used.

Chapter Review

Chapter Review

ALTERNATIVE ASSESSMENT

The *Prentice Hall Science* program includes a variety of testing components and methodologies. Aside from the Chapter Review questions, you may opt to use the Chapter Test or the Computer Test Bank Test in your *Test Book* for assessment of important facts and concepts. In addition, Performance-Based Tests are included in your *Test Book*. These Performance-Based Tests are designed to test science process skills, rather than factual content recall. Since they are not content dependent, Performance-Based Tests can be distributed after students complete a chapter or after they complete the entire textbook.

CONTENT REVIEW

Multiple Choice

  1. a
  2. d
  3. d
  4. a
  5. c
  6. b
  7. c
  8. a

True or False

  1. F, hard
  2. F, mold
  3. T
  4. F, horizontal
  5. F, absolute age
  6. T
  7. T

Concept Mapping

  Row 1: Petrified, Mold, Cast, Imprint, Trace
  Row 2: Freezing, Amber, Tar pits

Content Review

Multiple Choice

Choose the letter of the answer that best completes each statement.

1. The shape of an organism preserved in rock is called a(an)
  a. mold.   c. imprint.
  b. cast.   d. petrification.

2. Bodies of whole animals have been preserved in
  a. ice.   c. amber.
  b. tar.   d. all of these

3. Rocks formed from the piling up of layers of dust, dirt, and sand are called
  a. igneous.   c. magma.
  b. metamorphic. d. sedimentary.

4. A crack in a rock structure that moves the rocks on either side out of line is a(an)
  a. fault.   c. intrusion.
  b. cast.   d. extrusion.

5. The decay rate of a radioactive element is measured by a unit called
  a. period.   c. half-life.
  b. era.   d. unconformity.

6. Dinosaurs found at Bone Cabin Quarry lived during the
  a. Paleozoic Era. c. Cretaceous Period.
  b. Jurassic Period. d. Tertiary Period.

7. The animal used as an index fossil for the Paleozoic Era is the
  a. sago palm.   c. trilobite.
  b. dinosaur.   d. *Eryops*.

8. A measure of how many years ago an event occurred or an organism lives is
  a. absolute age. c. decay time.
  b. relative age.   d. sedimentary age.

True or False

If the statement is true, write "true." If it is false, change the underlined word or words to make the statement true.

1. The <u>soft</u> parts of plants or animals usually become fossils.

2. An empty space called a <u>cast</u> is left in a rock when a buried organism dissolves.

3. Footprints of extinct dinosaurs are examples of <u>trace fossils</u>.

4. Sediments are usually deposited in <u>vertical</u> layers.

5. The measure of how many years ago an event occurred or an animal lived is called <u>relative age</u>.

6. <u>Faults</u> are always younger than the rock layers they cut through.

7. The time it takes for half the atoms in a sample of a radioactive element to decay is called its <u>half-life</u>.

Concept Mapping

Complete the following concept map for Section 1–1. Refer to pages F6–F7 to construct a concept map for the entire chapter.

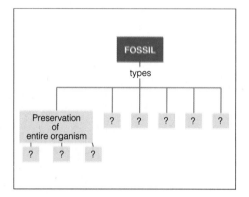

CONCEPT MASTERY

1. The breakdown of a radioactive element into a decay element occurs at a constant rate. By measuring the amount of the decay element contained in a fossil, the age of the fossil can be estimated.

2. Volcanoes and mountains formed, first fish appeared, coral reefs formed, first land plants and air-breathing animals developed, first forests grew, and amphibians, sharks, and insects developed.

3. The rate of deposition of sediment is not constant.

4. Most fossils are found in sedimentary rock because the sediments covering the organism cause a relatively quick burial that slows down decay. Quick burial in sediments also prevents animals from eating the dead organism. The forces that create igneous and metamorphic rocks would also be destructive to the remains of most organisms.

5. Fossils can form by means of petrification, molds and casts, imprints, preservation of whole animals, and trace fossils.

Concept Mastery

Discuss each of the following in a brief paragraph.

1. How are radioactive elements used to determine the age of rocks and fossils?
2. Describe some of the geologic changes that occurred during the Paleozoic Era.
3. Why is sedimentation rate an inaccurate way of measuring geologic time?
4. Why are few fossils found in igneous or metamorphic rock?
5. Discuss five ways fossils can form.
6. Trilobites are important index fossils for the Paleozoic Era. Explain what is meant by this statement.

Critical Thinking and Problem Solving

Use the skills you have developed in this chapter to answer each of the following.

1. **Making calculations** A radioactive element has a half-life of 500 million years. After 2 billion years, how many half-lives have passed? How many kilograms of a 10-kilogram sample would be left at this time? If the half-life were 4 billion years?
2. **Analyzing diagrams** Use the diagram to answer the following questions.
   a. According to the way in which layers C, D, E, and F lie, what might have happened in the past?
   b. Which letter shows an unconformity? Explain your answer.
   c. List the events that occurred from oldest to youngest. Include the order in which each layer was deposited and when the fault, intrusion, and unconformity were formed. Explain why you chose this order.

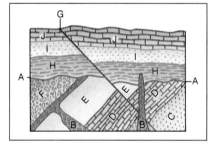

3. **Interpreting evidence** Suppose you are a scientist who finds some fossils while looking at a cross section of rock in an area. One layer of rock has fossils of the extinct woolly mammoth. In a layer of rock below this, you discover the fossils of an extinct alligator. What can you determine about changes over time in the climate of this area?
4. **Developing a theory** Explain why an animal species that reproduces every year would have a better chance of surviving a change in its environment than an animal species that reproduces only once or twice in ten years does.
5. **Sequencing events** List five events in your life in the order they happened. Have a friend or classmate list five events in his or her life in the order they happened. Now try to list all ten events in the order they happened.
   What difficulties did you have in deciding whether a certain event occurred before or after other events? How has this activity helped you understand the difficulty scientists had in developing a time scale without radioactive dating?
6. **Using the writing process** Choose one of the four geologic eras and write a short story depicting a day in the life of an organism living during that era.

KEEPING A PORTFOLIO

You might want to assign some of the Concept Mastery and Critical Thinking and Problem Solving questions as homework and have students include their responses to unassigned questions in their portfolio. Students should be encouraged to include both the question and the answer in their portfolio.

ISSUES IN SCIENCE

The following issues can be used as springboards for discussion or given as writing assignments.
1. Do you think the division of geologic time into the specific eras and periods presently in use will always be the one that is used? If not, what might bring about a change in the system used?
2. Not all scientists agree on the age of the Earth. Find out why such controversy exists; then state your opinion on the issue.

---

6. The trilobites were widely distributed and evolved rapidly during the Paleozoic Era and had different forms in each of the era's periods. Thus, their fossils can be used to help date rock layers of that era.

CRITICAL THINKING AND PROBLEM SOLVING

1. After 2 billion years, four half-lives have passed. At that time, 0.625 kg would be left. If the half-life were 4 billion years, there would be 7.5 kg left because only one half of one half-life would have passed. Students should multiply 10 kg by 0.75 to get 75 percent.
2. In this diagram, students should note that the rock layers were first tilted and then eroded. Layers from oldest to youngest are C, D, E (all deposited), F, A (unconformity), H, I, J (all deposited), G (fault occurred), B (intrusion).
3. The alligator lives in a wet, swampy area. The woolly mammoth lived in cold regions often covered by ice with just traces of plants. Thus, the climate of the region changed from a warm, wet area to a cold, barren area.

4. Accept all logical responses. The faster a population produces new generations, the more quickly adaptations can occur.
5. Pairs of students will probabiy have difficulty listing all ten events in order. They should realize that developing a time scale requires a series of events that take place at a steady rate.
6. Accept all logical, well-written responses. Students' stories should be consistent with the information presented in the chapter.

Chapter 2 CHANGES IN LIVING THINGS OVER TIME

SECTION	HANDS-ON ACTIVITIES
2–1 Evolution: Change Over Time pages F48–F56 Multicultural Opportunity 2–1, p. F48 ESL Strategy, 2–1, p. F48	**Student Edition** ACTIVITY (Doing): Extinct Species, p. F55 LABORATORY INVESTIGATION: Analyzing a Geologic Time Line, p. F72 **Activity Book** CHAPTER DISCOVERY: Natural Selection, p. F55 **Teacher Edition** Mechanical Evolution, p. F46d
2–2 Charles Darwin and Natural Selection pages F57–F61 Multicultural Opportunity, 2–2, p. F57 ESL Strategy, 2–2, p. F57	**Student Edition** ACTIVITY (Discovering): Survival of the Fittest, p. F59 ACTIVITY BANK: Where Are They? p. F106 ACTIVITY BANK: Variety Is the Spice of Life, p. F108 **Laboratory Manual** Variation and Natural Selection, p. F29 Variation in a Classroom Population, p. F35
2–3 The Development of a New Species pages F62–F68 Multicultural Opportunity 2–3, p. F62 ESL Strategy 2–3, p. F62	**Student Edition** ACTIVITY (Doing): Living Dinosaurs? p. F67 **Teacher Edition** Adaptation Through Human Intervention, p. F46d
2–4 Punctuated Equilibrium pages F69–F71 Multicultural Opportunity 2–4, p. F69 ESL Strategy 2–4, p. F69	
Chapter Review pages F72–F75	

OUTSIDE TEACHER RESOURCES

Books

Dinley, David. *Earth's Voyage Through Time,* Knopf.

Gallant, Roy A. *Before the Sun Dies: The Story of Evolution,* Macmillan.

Haber, Francis C. *The Age of the World: Moses to Darwin,* Greenwood.

Jacobson, Claire. *The Hunters of Prehistory,* Atheneum.

Livingstone, David N. *Darwin's Forgotten Defenders,* Eerdmans.

Taylor, Ron. *The Story of Evolution,* Warwick.

OTHER ACTIVITIES	MEDIA AND TECHNOLOGY
Student Edition ACTIVITY (Writing): A Packy-Poem, p. F49 **Activity Book** ACTIVITY: Interpreting the Geologic Time Scale, p. F67 **Review and Reinforcement Guide** Section 2–1, p. F13	**Video/Videodisc** Patterns of Evolution Patterns of Diversity (Supplemental) **Courseware** Food Chains and Food Webs (Supplemental) **English/Spanish Audiotapes** Section 2–1
Activity Book ACTIVITY: Mermaids on the Endangered Species List, p. F59 ACTIVITY: Modifying the Manatee, p. F61 **Review and Reinforcement Guide** Section 2–2, p. F17	**Interactive Videodisc/CD ROM** Paul ParkRanger and the Mystery of the Disappearing Ducks **Interactive Videodisc** Insects: Little Giants of the Earth **Courseware** Adaptation in Two Biomes (Supplemental) **English/Spanish Audiotapes** Section 2–2
Student Edition ACTIVITY (Writing): Warmblooded Dinosaurs, p. F63 **Activity Book** ACTIVITY: Adaptive Radiation, p. F63 ACTIVITY: Design a Bird, p. F65 **Review and Reinforcement Guide** Section 2–3, p. F19	**Interactive Videodisc/CD ROM** Virtual BioPark Amazonia **English/Spanish Audiotapes** Section 2–3
Review and Reinforcement Guide Section 2–4, p. F23	**English/Spanish Audiotapes** Section 2–4
Test Book Chapter Test, p. F31 Performance-Based Tests, p. F71	**Test Book** Computer Test Bank Test, p. F39

*All materials in the Chapter Planning Guide Grid are available as part of the Prentice Hall Science Learning System.

Audiovisuals

Darwin and Diversity, video, The Media Guild

Earth: Treasures of the Earth, filmstrip with cassette, Prentice-Hall Media

Glaciers and the Ice-Age, filmstrips with cassettes, Encyclopaedia Britannica Education

The Old Stone Age, filmstrip with cassette, Society of Visual Education

Prehistoric Life: The Origins, filmstrips with cassettes, Society of Visual Education

Time of Man, Parts 1 and 2, 16-mm film, BFA

CHAPTER OVERVIEW

A change in a species over a period of time is called evolution. The theory of evolution explains how living things change. The theory of evolution has been tested, retested, developed, and revised for over a century. In the early 1800s, a French biologist named Jean-Baptiste de Lamarck began to develop some theories about evolution based on anatomy. Since that time, more and more supporting evidence has led to its usefulness in explaining the history of living things.

Fossils are imprints or remains of plants or animals that existed in the past. Fossil evidence supplies a record of the history of living things. Modern techniques of radioactive dating and chemical analysis give further evidence in support of the theory of evolution.

Nature's demands on living things tend to select the "fittest" of the species to survive. Natural selection is the survival and reproduction of those organisms best adapted to their surroundings.

Although the theory of natural selection as an explanation for evolution is accepted by most scientists, some believe that the process occurs gradually and at a relatively constant rate. Other scientists, in contrast, feel that evolution occurs in a series of rapid changes. The former theory is called gradualism; the latter is referred to as punctuated equilibrium.

2–1 EVOLUTION: CHANGE OVER TIME
THEMATIC FOCUS

The purpose of this section is to introduce students to the theory of evolution. Evolution is a change in a species over time. As the face of the Earth changed, many organisms died because they could not compete with other organisms. Mutations of genes produced new or slightly modified organisms that survived and reproduced because they were able to meet the demands of their environment.

There are many different types of evidence that support the theory of evolution. The four major types of evidence are anatomical, fossil, embryological, and chemical.

The themes that can be focused on in this section are evolution, patterns of change, and scale and structure.

*Evolution: Evolution is a change in species over time.

*Patterns of change: Mutations are the driving force behind evolution. Mutations that increase an organism's chances for survival are called adaptations.

Scale and structure: A molecular clock can be used to demonstrate how closely related two species are and estimate how long ago they separated from a common ancestor.

PERFORMANCE OBJECTIVES 2–1

1. Define evolution.

2. Explain how an adaptation can increase an organism's chance for survival.

3. Describe the chemical, anatomical, and fossil evidence for evolution.

SCIENCE TERMS 2–1

evolution p. F48
adaptation p. F49
homologous structure p. F53
molecular clock p. F55

2–2 CHARLES DARWIN AND NATURAL SELECTION
THEMATIC FOCUS

The purpose of this section is to introduce students to Darwin's theory of natural selection. Natural selection is the survival and reproduction of those organisms best adapted to their surroundings. Charles Darwin theorized that organisms that could adapt to their environment and pass these adaptations on to their offspring would best survive. The process in which only the best-adapted members of a species survive is often referred to as "survival of the fittest."

Members of the same species vary in many slight ways. These differences, or variations, can lead to new species over long periods of time. Many times these new species have a better chance for survival than the original species.

The themes that can be focused on in this section are unity and diversity and stability.

*Unity and diversity: Although all members of a species share similar characteristics and can interbreed, they do exhibit minor variations from one another.

Stability: Natural selection is the survival and reproduction of those organisms best adapted to their environments.

PERFORMANCE OBJECTIVES 2–2

1. Define natural selection.
2. Determine the effects of variation on natural selection.

SCIENCE TERMS 2–2

natural selection p. F58

2–3 THE DEVELOPMENT OF A NEW SPECIES
THEMATIC FOCUS

The purpose of this section is to show students the effects of migration and isolation on the process of evolution. Migration is defined as the moving away of an organism from its original home to a new place. Isolation is defined as the separation of some members of a species or group of species from the rest of their kind for long periods of time. The isolation of many Australian species is given as an example.

Both migration and isolation can lead to the development of new species, as part of an original population occupies a new niche and starts to adapt to its new environment.

The themes that can be focused on in this section are systems and interactions and patterns of change.

*Systems and interactions: When two species share the same niche, they compete for resources. Often, one of the species becomes extinct.

*Patterns of change: Adaptive radiation is the process by which one species evolves into several species, each filling a different niche.

PERFORMANCE OBJECTIVES 2–3

1. Explain how natural selection leads to new and varied species.

2. Define migration and discuss its effect on evolution.

3. Define isolation and discuss its effect on evolution.

SCIENCE TERMS 2–3

niche p. F62

adaptive radiation p. F67

2–4 PUNCTUATED EQUILIBRIUM
THEMATIC FOCUS

The purpose of this section is to show students two differences between two of the theories of evolution.

Charles Darwin believed in gradualism, or a slow and steady process of evolution. Certain evidence in the fossil record, however, indicates that periods of rapid evolutionary change may also have occurred. Stephen Gould and Niles Eldridge have developed a theory known as punctuated

equilibrium. This theory supports the findings of fossil records. It states that a species may have little or no change over a long period of time. Then there is a rapid and sudden change resulting in a new species.

Many scientists believe that the combination of slow and rapid evolution is responsible for the plant and animal species found on the Earth today.

The themes that can be focused on in this section are evolution and patterns of change.

*Evolution: All living things have evolved through natural selection from other living things.

*Patterns of change: Evolution may be gradual, as described by Darwin, or more rapid, as described by Gould and Eldridge.

PERFORMANCE OBJECTIVES 2–4

1. Explain the gradualism theory of natural selection.

2. Explain the punctuated equilibrium theory of natural selection.

SCIENCE TERMS 2–4

punctuated equilibrium p. F69

Discovery *Learning*

TEACHER DEMONSTRATIONS MODELING

Mechanical Evolution

Show students a photo of one of the first automobiles and a modern car (or early airplane and a modern jet).
- **How are these two automobiles (airplanes) similar?** (Accept all logical answers.)
- **How are these two automobiles (airplanes) different?** (Accept all logical answers.)
- **How did the automobile (airplane) change over the years?** (Accept all logical answers.)

- **Why do you think the engineers changed the automobile (airplane)?** (Accept all logical answers.)
- **Was the change planned or was it accidental?** (Accept all logical answers.)

Adaptation Through Human Intervention

Show students photographs or illustrations of cattle, horses, hogs, corn, and other domestic animals and plants as they looked in the late 1800s. Compare these with recent pictures of the different breeds and varieties found on today's ranches and farms. Have students note the changes found in organisms in recent times.
- **How is the modern-day organism different from its ancestor?** (Answers will vary.)
- **How are these differences of value to us today?** (Answers will vary.)

Changes in Living Things Over Time

INTEGRATING SCIENCE

This life science chapter provides you with numerous opportunities to integrate other areas of science, as well as other disciplines, into your curriculum. Blue numbered annotations on the student page and integration notes on the teacher wraparound pages alert you to areas of possible integration.

In this chapter, you can integrate language arts (pp. 48, 49, 55, 63), life science and classification (p. 49), life science and genetics (p. 49), life science and zoology (p. 50), life science and DNA (p. 54), life science and proteins (p. 55), geography (p. 57), social studies (pp. 57, 61), life science and seed dispersal (p. 58), life science and metamorphosis (p. 59), life science and ecology (p. 62), earth science and continental drift (p. 64), life science and bird adaptations (p. 65), fine arts (p. 67), and earth science and the solar system (p. 71).

SCIENCE, TECHNOLOGY, AND SOCIETY/COOPERATIVE LEARNING

A species has traditionally been defined as a group of organisms that does not breed and produce fertile offspring with other groups of organisms. It can happen, however, that two species "break the rules" and hybridize as an evolutionary response to environmental disturbance. This idea challenges the definition of a species and presents challenges as to how species should be protected if threatened.

The wolf-coyote mating is an example of this phenomenon. The wolf population has been drastically reduced by human activities. When male wolves searching for hard-to-find females encountered receptive female coyotes, mating occurred and the hybrid red wolf was "born." Red wolves, by definition, are a species—they breed only with other red wolves.

Hybridization may create a precarious legal position for these new species. Be-

INTRODUCING CHAPTER 2

DISCOVERY LEARNING

▶ *Activity Book*

Begin your introduction to this chapter by using the Chapter 2 Discovery Activity from your *Activity Book*. Using this activity, students will discover how the process of natural selection might occur in mice.

USING THE TEXTBOOK

Begin your introduction of this chapter by using a rough time line to illustrate the history of the Earth. Explain that all the events that have occurred on Earth can be better understood if they are viewed as having taken place in the span of a year. Using this scale, each day would represent 12.3 million years in the Earth's history.

Sketch a time line on posterboard or the chalkboard, making one day equal to

Changes in Living Things Over Time

Guide for Reading

After you read the following sections, you will be able to

2–1 Evolution: Change Over Time
- Describe evidence of evolution.

2–2 Charles Darwin and Natural Selection
- Define and describe natural selection.

2–3 The Development of a New Species
- Describe the processes of speciation and adaptive radiation.

2–4 Punctuated Equilibrium
- Describe punctuated equilibrium and relate it to adaptive radiation.

In the fall of 1991, scientists from Montana State University, the University of Wyoming, and the Royal Tyrell Museum of Canada quickly prepared to leave on an unusual rescue mission. No, they weren't going to rescue an injured camper. Nor were they trying to save an animal in danger. In fact, they were going to rescue a creature that had died about 150 million years ago!

The scientists were on their way to northern Wyoming, where the first intact skeleton of a dinosaur called *Allosaurus* had been discovered. Until then, only bits and pieces of *Allosaurus* bones had been found, but never a complete skeleton. The rescue mission involved preserving the find before the harsh Wyoming winter set in.

Even as you read this chapter, other fossils that provide clues to Earth's past are being sought. The Earth gives up its secrets grudgingly, one clue at a time. Do we have all the answers? Not by a long shot. But we do know a good deal about the changes that have occurred in living things over time—changes we call the evolution of living things.

Journal *Activity*

You and Your World Suppose you were lucky enough to discover the bones of a creature that walked the Earth millions of years ago. In your journal, describe your thoughts and feelings about finding an organism never before seen by people. If you wish, include a drawing of your find.

◀ *Allosaurus, which lived about 140 million years ago, was one of the fiercest predators ever to walk the Earth. It was also one of the largest—about 8 meters long!*

F ■ 47

---

cause the Endangered Species Act uses the traditional definition for species, hybrid organisms are excluded from protection. The red wolf is one of the "hybrid species" most affected—it was placed on the endangered species list before scientists discovered that it was a hybrid.

Cooperative learning: Using preassigned lab groups or randomly selected teams, have groups complete one of the following assignments.

- Invent a hybrid species that will be better able to survive in a habitat that is changing due to human activities. The final product of each group should be a series of labeled diagrams and pictures that includes a description of the original species and its habitat, the new habitat and the cause of change, the hybrid species, and an explanation of how the hybrid is better adapted to the new habitat.
- Write a letter to their congressional representative explaining why or why not they think the Endangered Species Act should be changed to include hybrid species.

See Cooperative Learning in the *Teacher's Desk Reference.*

JOURNAL ACTIVITY

You may want to use the Journal Activity as the basis of class discussion. Before students begin their journal entries, point out that archaeologists often find other interesting artifacts with fossilized bones. Students might discuss other finds such as plants and human-made tools in their entries. Students should be instructed to keep their Journal Activity in their portfolio.

---

a 1-cm length. The total length of the time line will be 365 cm, representing one year.

Show students that the first forms of life would appear at approximately 120 cm, the first dinosaurs at approximately 334 cm, the first mammals at 349 cm and the first modern humans at 364.9 cm. Explain to students that the Earth and its animals and plants have changed many times in the course of history. Then have them examine the illustration on page 46.

- **What information do people use to create drawings such as the drawing on page 46?** (Information gained from the fossil record and deposited rock layers helps people depict life as it was millions of years ago.)
- **Why might these drawings change over the years?** (As more fossil evidence is uncovered, the re-creations are changed and updated.)

Then have students read the text on page 47.

- **Why would a complete fossil skeleton of a dinosaur create excitement among scientists?** (A complete skeleton would enable scientists to better guess at the structure and function of the dinosaur's organs. And it would help scientists understand the behavior of the dinosaurs.)
- **Why has it been difficult for scientists to reconstruct the early history of the Earth?** (Accept all logical responses. Students may mention that the fossil record is incomplete.)

2-1 Evolution: Change Over Time

Guide for Reading

Focus on these questions as you read.
▶ What is evolution?
▶ How are mutations related to the evolution of living things?

2-1 Evolution: Change Over Time

In Chapter 1 you read about Earth's history. You learned that by studying rocks and fossils, scientists have developed a fairly accurate picture of how Earth and its inhabitants have changed over time. Although the picture is far from complete, there is no doubt that changes have occurred and that many of the living things on Earth today are very different from the living things that existed in the past. In other words, there is no doubt that living things have changed over time.

How and why have living things changed? And which living things are more closely related to one another? Today scientists know that the answers to these questions lie in the process of **evolution.** The word evolution comes from Latin and means an unfolding or opening out. A scientific translation of this meaning is descent with modification. Descent means to come from something that lived before. And modification means change. Thus evolution means that all inhabitants of Earth are changed forms of living things that came before.

Evolution can be defined as a change in species over time. A species is a group of organisms that share similar characteristics and that can interbreed with one another to produce fertile offspring. Lions,

Figure 2-1 *The members of the species* Alces alces, *better known as moose, often wade into lakes and ponds to graze on water plants. What is a species?* ❶

for example, are a species. So are tigers. Lions and tigers share many similar characteristics and can even be bred together to produce offspring called ligers and tiglons. The offspring, however, are not fertile. That is, ligers and tiglons cannot mate and produce more of their own kind. So lions and tigers are not the same species. They are two separate species. Quite the opposite is true of a German shepherd and a French poodle. Although they appear quite different, they can interbreed and produce fertile offspring. So dogs, even though they may appear quite different, are all members of the same species.

Why have some species evolved into the plants and animals living on Earth today while other species became extinct? During the history of life on Earth, chance changes in the genes of organisms have produced new or slightly modified living things. A gene is a unit of heredity that is passed on from parent to offspring. A change in a gene will produce a change in the offspring of an organism. Changes in genes are called mutations. And mutations are one of the driving forces behind evolution.

Mutations: Agents of Change

Most of the time, a mutation in a gene produces an organism that cannot compete with other organisms. This new organism usually dies off quickly. Sometimes, however, the change in the organism is a positive one. The change makes the organism better suited to its environment. A change that increases an organism's chances of survival is called an **adaptation.**

Organisms that are better adapted to their environment do more than just survive. They are able to produce offspring, which produce more offspring, and so on. Over a long period of time, so many small adaptations may occur that a new species may evolve. The new species may no longer resemble its ancient ancestors. In addition, the new species may be so successful in its environment that the species from which it evolved can no longer compete. The original species dies off. Thus the development of a new species can result in the extinction of another species.

ACTIVITY READING

A Packy-Poem

Ever wonder how poets view those "terrible lizards" we call dinosaurs. For a humorous point of view, read the poem *Pachycephalosaurus* by Richard Armour.

F ■ 49

ACTIVITY READING
A PACKY-POEM

Skill: Reading comprehension

In this activity, students will see how one writer found a practical application for scientific ideas. After students have read the poem by Richard Armour, challenge them to write dinosaur poems of their own. Or they might choose to write poems about evolution, mutations, or adaptations.

Integration: Use this Activity to integrate language arts into your lesson.

● ● ● ● **Integration** ● ● ● ●

Use the discussion of evolution to integrate concepts of language arts into your lesson.

CONTENT DEVELOPMENT

Point out that we can recognize similarities between past organisms and some of the organisms living today. Tell students that when heritable traits appear in a new organism, they are usually due to mutations.

● ● ● ● **Integration** ● ● ● ●

Use the discussion of lions and tigers to integrate concepts of classification into your lesson.

CONTENT DEVELOPMENT

Point out that a new plant or animal species may first come into being as a result of mutation. Explain that if the animal or plant can survive in the environment, it will pass the trait to its offspring. Tell students that a change that increases an organism's chances of survival is known as an adaptation. Point out that if a certain type of plant or animal cannot survive and dies completely, we refer to it as extinct.

● ● ● ● **Integration** ● ● ● ●

Use the discussion of genes to integrate concepts of heredity into your lesson.

REINFORCEMENT/RETEACHING

Review the term *adaptation* to ensure that all students understand this important concept in evolution.

Although carbon-14 dating is often featured in textbooks, it is useless for dating most of the fossils described. Besides the very short half-life of this element, living organisms take up radioactive carbon-14 from their environment along with non-radioactive carbon.

Figure 2–2 *A hummingbird's beak, a giraffe's neck, and a vampire bat's sharp teeth are all examples of adaptations for feeding. What are adaptations? How do these particular adaptations affect the animals that have them?* ❶

The Fossil Record

In Chapter 1 you read about the many types of fossils that have been found. The record of Earth's history in fossils and in fossil-containing rocks clearly demonstrates that living things have evolved, or changed over time. You can see some of these changes for yourself in Figure 2–3, which shows the fossil record of the camel. Scientists have cataloged the evolution of many organisms, such as the camel.

The fossil record provides evidence about the changes that have occurred in living things and their way of life. In 1983, for example, scientists found a buried skull belonging to an animal that had lived more than 50 million years ago. The skull was very similar to that of a whale. But the bony structure ❶ that allowed the animal to hear could not have worked underwater. So scientists concluded that the whalelike skull belonged to an ancestor of modern whales that spent some of its life on land. Scientists also concluded that at some point the ancestors of whales left the land completely and became water-dwelling creatures.

Although the fossil record is not complete—and never will be because many organisms have come and gone without leaving any fossils—it does provide

50 ■ F

2–1 (continued)

REINFORCEMENT/RETEACHING

Stress that fossils are normally found in sedimentary rock that is formed by the gradual hardening of layers of mud, sand, or clay. If a dead organism is covered immediately by sediments, it is protected from decomposition. As years pass, more and more mud, sand, or clay is deposited on top of the organism. After hundreds of thousands of years, the tremendous weight of the top layers of sediment cause the bottom layers to change to sedimentary rock.

Fossils can take many different forms. If an organism eventually decays after the sediments change to rock, its imprint will be left in the rock. This outline, called a mold, can become filled with minerals. When this happens, it is called a cast. If the soft tissues of the organism are slowly replaced by minerals, the organism is said to be petrified. Also, as mentioned before, whole organisms can be preserved in materials such as ice, tar, and amber.

CONTENT DEVELOPMENT

Because the lowest layers of rock are the oldest, scientists can estimate the age of a fossil by noting its placement in a rock layer. Stress that rock formations, such as the Grand Canyon, provide scientists with many varieties of fossils from many different geologic time periods. Mention that the Grand Canyon contains fossils ranging in size and age from single-celled algae to large dinosaurs and trees.

Emphasize that all of the fossil evidence that scientists have collected is known as the fossil record. Stress that the fossil record is the most complete record of life on the Earth. It represents, however, only a fraction of the total number of life forms found throughout time. Most living things die under conditions that are not favorable for the preservation of their remains. As a result, they have disappeared from the Earth without leaving a trace.

● ● ● ● **Integration** ● ● ● ●

Use the discussion of the fossil record to integrate concepts of zoology into your lesson.

ENRICHMENT

Tell students that fossils must be dated to be of any value. Explain that both radioactive dating and relative dating can be used to date fossils.

Point out that radioactive dating is the process of using radioactive elements.

Figure 2–3 *Living things have evolved, or changed over time. The diagram shows the fossil record of the past 65 million years in the evolution of the camel. How are modern camels similar to their ancestors? How are they different?* ❷

The skulls shown are labeled, from top to bottom: Pleistocene, Pliocene, Miocene, Oligocene, Eocene, Paleocene.

ample evidence that evolution has indeed occurred. Fortunately, fossils are but one piece of evidence of evolution. Now let's look into some other ways scientists explore the evolution of living things.

Anatomical Evidence of Change

In the early 1800s, a French biologist named Jean-Baptiste de Lamarck came to the conclusion that living things had changed over time. In his book *Philosophie zoologique,* Lamarck suggested that species seemingly very different could be proven by

BACKGROUND INFORMATION
LAMARCKISM

One of the earliest theories of evolution was proposed by the French biologist Jean-Baptiste de Lamarck (1744–1829). Lamarck's theory of evolution involved two main principles: (1) the theory of use and disuse and (2) the theory of inheritance of acquired characteristics. According to the theory of use and disuse, the more an organism uses a particular part of its body, the stronger and better developed that part becomes. The less a part is used, the weaker and less developed it becomes. The theory of the inheritance of acquired characteristics states that the physical characteristics an organism develops through the use or disuse of its different body parts could be passed on to its offspring.

Explain that a radioactive element gives off radiation as the atoms of the element break down into different elements. Each radioactive element decays or breaks down at a different rate. This difference in rates is used to determine the age of the fossil. Explain that the rate of decay is measured in a unit called a half-life.

Point out another way of dating fossils is relative dating. Explain that rocks and fossils are deposited "in sequence" in sedimentary rocks. The oldest sediments

were deposited on the bottom first, along with the fossils of that time period. Then younger sediments with more recent fossils were deposited on top of the oldest layer. The oldest sedimentary rocks and fossils are therefore at the bottom. And younger sedimentary rocks and fossils are on top. With this information, the relative age of the rock formation and fossils can be determined.

GUIDED PRACTICE

Skills Development

Skills: Applying concepts, making observations, interpreting charts

At this point have students complete the in-text Chapter 2 Laboratory Investigation: Analyzing a Geologic Time Line. In the investigation, students will plot evolutionary events on a time line.

In the eighteenth and nineteenth centuries, many people believed in the inheritance of acquired characteristics; the notion is not unique to Lamarck. People later focused on this particular notion, however, to the exclusion of the rest of Lamarck's theory, and "Lamarckian evolution" became equated with the inheritance of acquired characteristics.

If a change in an organism's habits or environment causes a structure to become unnecessary, natural selection will no longer work to keep that structure in the population. If a mutation occurs that eliminates the structure without otherwise harming the organism, the structure can disappear.

But remember that genetic variation is random. Just because a structure is not used in an animal does not mean that there will be mutations to get rid of that structure. It is likely, however, that there is genetic variation in the size of that structure. Genes that cause a useless structure to get smaller will certainly not hurt the organism. Those genes may even help the organism if they avoid wasting energy in growing structures that serve no purpose. For this reason genetic changes that cause useless structures to get smaller can take over the population.

2–1 (continued)

REINFORCEMENT/RETEACHING

Explain that fossil evidence can also be used to support the theory of evolution. By comparing the fossilized skull of a 50-million-year-old whale to the skull of a modern-day whale, scientists have theorized that at one time a whalelike animal was able to live on the land. As time passed, this organism evolved into a water-dwelling organism.

Stress that fossils are the imprints or remains of a plant or animal that lived in

the past. Point out that most fossils are actually a part of an organism or an imprint from a part of an organism. In some cases an organism dies under unusual conditions and its entire body is preserved. Examples of this are insects encased in amber, animals trapped in the La Brea tar pits in California, and woolly mammoths frozen in ice.

FOCUS/MOTIVATION

Have students observe Figure 2–4. Discuss the pictures with some of the following questions.
• **What do you observe in Figure 2–4?** (Accept all logical answers.)
• **What do the organisms have in common?** (Accept all logical answers.)

Have students focus in on the bone structures.
• **What do you see that is similar about the bone structures?** (Accept all logical answers.)

Figure 2–4 Although Lamarck was one of the first scientists to recognize that evolution occurred, his theories about how and why evolution took place proved to be incorrect. How would Lamarck have explained the evolution of the ostrich's strong legs for running? The snake's lack of limbs? How would a modern scientist explain their evolution? ❶

close study to have developed from the same ancestors. "All forms of life could be organized into one vast family tree," he said. Lamarck's writings were a milestone in biology, as he was one of the first scientists to recognize that evolution had occurred. At the time, Lamarck's theories were contrary to popular beliefs.

Despite his insight into the concept of evolution, Lamarck proved to be wrong about most of his theories concerning the process of evolution. Lamarck believed that organisms change because of an inborn will to change. He believed, for example, that the ancestors of birds had a desire to fly. Over many years, that desire enabled birds to acquire wings—and to be better adapted to their environment as well.

Lamarck also believed that organisms could change their body structure by using body parts in new ways. For example, because birds tried so hard to use their front limbs for flying, the limbs eventually changed into wings. In much the same way, Lamarck reasoned, a body part would eventually grow smaller or even disappear from disuse. For example, by slithering along the ground, a snake would eventually lose its limbs.

As you have read earlier, one of the driving forces behind evolution are mutations. And mutations are changes in genes. Lamarck's beliefs about how and why organisms change are incorrect. Wanting a new body part cannot cause a mutation. Using a body part in a different way—or not using it at all—cannot cause a mutation. Mutations happen; they are chance events. Mutations are independent of a desire to change and of a need to adapt to the environment.

Although Lamarck failed to explain the mechanics of evolution, he did provide a new way of studying living things. All of Lamarck's theories were based on the evidence of anatomy. Anatomy is the study of the physical structure of living things.

Look closely at the structure of the bones in the bat's wing, dog's foreleg, whale's fin, and human's arm in Figure 2–5. Do you see any similarities in the shapes and arrangement of the bones of these animals? You should, because similarities do exist. ❷ These similarities indicate that each organism has evolved from a common ancestor. The whale became

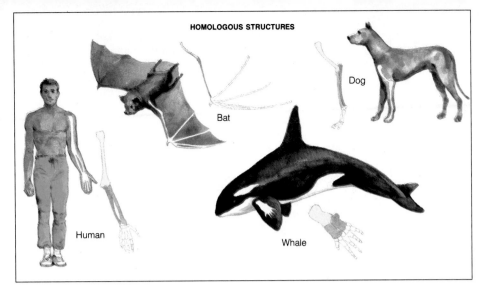

HOMOLOGOUS STRUCTURES

Human

Bat

Dog

Whale

Figure 2–5 *Human arms, bat wings, whale flippers, and dog legs are superbly adapted to performing different tasks. However, their internal structure is remarkably similar. What are such structures called? What do they indicate?* ❸

better adapted to an ocean environment as the structure of its bones gradually changed to those found in the fins of a modern whale. Similarly, the bones of a bat's forelimbs gradually evolved into those found in the wings of modern bats. Structures such as these that evolve from the same ancestral body parts are called **homologous** (hoh-MAH-luh-guhs) **structures.** It was through the study of homologous structures that Lamarck concluded that living things had evolved and had become better adapted to their environment.

Embryological Evidence of Change

By the end of the nineteenth century, scientists had observed that the embryos of many different animals appeared so similar that it was difficult to tell them apart. An embryo is an organism in its early stages of development. Look at Figure 2–6 on page 54. As you can see, the embryos of a fish, chicken, rabbit, and human appear almost identical in their early stages. What does this mean? ❹

The growth and development of an embryo is controlled by its genes. Similarities in the embryos of different organisms indicate that these organisms share a common ancestor. That is, these organisms all share a common heritage. As the embryos

F ■ 53

GUIDED PRACTICE

Skills Development

Skill: Making comparisons

Obtain samples of different bird wings—such as those of a chicken, turkey, quail, or dove—from a supermarket. Display the different wings in dissection trays and have students identify the homologous structures. Discuss the structure and function of a bird's wing. Based on their observations, have students decide if these organisms are closely related to each other.

CONTENT DEVELOPMENT

Explain that embryos of vertebrates, or animals with backbones, are very similar in the early stages. Explain that the more similar the structure of the embryos of different organisms, the more closely related those organisms are. Tell students that the study of embryos is called embryology. Point out that using Lamarck's theory of similar structures and embryology, scientists have added another piece of evidence to the theory of evolution.

CONTENT DEVELOPMENT

Point out that a French biologist named Jean-Baptiste de Lamarck suggested a theory of evolution based on the anatomy or structure of living things. Explain that he suggested that when body parts or organs are similar in structure in different organisms, the organisms are related. Tell students that when body parts or organs are similar in structure, they are said to be homologous.

ENRICHMENT

Show students photographs of organisms that have similar anatomies. Have students point out the similarities in structure. Stress that these similar body parts are said to be homologous. Make certain that students understand that scientists now believe that closely related organisms have homologous structures.

PROTEIN TAXONOMY

The more similar the proteins of two species are, the more closely related these species are likely to be. For example, virtually every organism has its own form of cytochrome *c*, a complicated protein molecule found in the electron transport chain. Differences in the various forms of cytochrome *c* are the result of mutations that occurred after the ancestors of the various species split apart. If two species diverged millions of years ago, there has been lots of time for mutations to alter the structure of cytochrome *c* molecules in each species. Analyses of the cytochrome *c* molecules have helped scientists determine that chickens and penguins, for example, are two closely related species.

2–1 (continued)

CONTENT DEVELOPMENT

Point out that chemical evidence is also used to support the theory of evolution.

Stress that the theory of evolution predicts that organisms that have similar physical characteristics will also have similarities in the structure of their DNA molecules. Remind students that DNA molecules make up the chromosomes that determine the organism's traits. Discuss the experiment conducted in 1984 by Dr. Wilson in which he found a 95 percent similarity in the structure of the DNA molecules of the quagga and the mountain zebra. You might want to mention that in 1985, upon further analysis of the quagga pelt, scientists discovered that the quagga was more closely related to the plains zebra than to the mountain zebra. Stress that scientists now use this method of DNA comparison to show the relationships between organisms.

Make certain that students understand that changes in organisms occur because of genetic mutations, or changes in the DNA structure. Remind students that although most mutations are harmful to organisms, some mutations allow the organism to better survive, passing on these genetic changes to future generations.

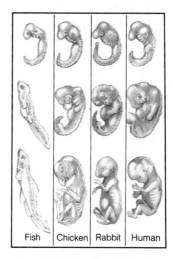

Figure 2–6 *The embryos of vastly different animals look quite similar during the earliest stages of development. These similarities in development hint at genes inherited from a common ancestor.*

Figure 2–7 *The DNA in a living zebra is 95 percent identical to that of the extinct quagga. This indicates that zebras and quaggas belong to separate but closely related species.*

Fish | Chicken | Rabbit | Human

continue to develop, other genes take over and the embryo becomes a fish, a chicken, a rabbit, or a human, depending on its gene structure. However, the similarities in the early stages of embryo development are further evidence that living things have evolved from earlier living things and that many living things share a common ancestor.

Chemical Evidence of Change

The quagga vanished from its home in South Africa about 100 years ago. Descriptions of the animal create a strange picture. The quagga had stripes like a zebra. But the stripes covered only its head, neck, and the front part of its body. Was this animal closely related to the zebra? Or was it a completely different kind of creature?

One type of evidence that might show if the zebra and the quagga evolved from a common ancestor is chemical similarities in their DNA. Earlier you read that genes are the units of heredity in living things. Genes are located on chromosomes, which ➊ are made of DNA. (The term DNA stands for deoxyribonucleic acid.)

From an evolutionary point of view, the more closely related two living things are, the more similar the structure of their DNA molecules will be. In 1985, Dr. Allen Wilson of the University of California at Berkeley analyzed DNA taken from muscle tissue of a preserved quagga. The tissue of the quagga had

● ● ● ● **Integration** ● ● ● ●

Use the discussion of chemical evidence of change to integrate concepts of DNA into your lesson.

FOCUS/MOTIVATION

Have students compare the pictures of the quagga and the zebra in Figure 2–7 as you describe the physical characteristics of the quagga.

• **What characteristics do the quagga and zebra have in common?** (The stripes cov-

ering the head, neck, and front of the body.)

• **How are the quagga and zebra different?** (The zebra's entire body is covered with stripes.)

• **Based on your observations, do these animals seem to be closely related?** (Students will most likely say that these organisms are closely related because they seem to have more similar characteristics than dissimilar ones.)

Stress to students that although some

been stored for more than a hundred years in a German museum. Dr. Wilson also analyzed DNA taken from a modern-day zebra. The structure of the DNA in the two samples was 95 percent identical.

Dr. Wilson concluded that the quagga and the zebra were, indeed, close relatives who shared a common ancestor about 3 million years ago. In other words, both the quagga and the zebra evolved from an animal that lived earlier. Scientists use this method of DNA comparison to show the relationships among many other types of organisms as well.

Molecular Evidence of Change

You have just read how similarities in DNA structure can be used to show relationships between organisms. Similarities in other kinds of molecules can be used in an almost identical way. Proteins are molecules that are used to build and repair body parts. Scientists believe that the more similar the structure of protein molecules of different organisms, the more closely related the organisms are. Scientists further believe that the closer the similarity in protein structure of different organisms, the more recently their common ancestor existed.

Scientists have developed a method of measuring the difference between the proteins of different species. In addition, scientists have developed a scale that can be used to estimate the rate of change in proteins over time. This scale of protein change is called a **molecular clock.**

By comparing the similarities in protein structure of different organisms, scientists can determine if the organisms have a common ancestor. If they do,

Figure 2–8 *Seventeen million years ago, a leaf fell from a tree and was perfectly preserved in mud. DNA studies revealed that the plant that dropped the leaf was related to the Chinese magnolia. By comparing DNA from the fossil leaf to DNA from the living plant, scientists can determine how fast DNA changes over time.*

ACTIVITY DOING

Extinct Species

Visit a museum of natural history. Find the exhibits of extinct animals such as dinosaurs, woolly mammoths, saber-toothed cats, and others. Find out when and where each of these animals lived. Also find out the reasons scientists believe these animals became extinct. Present your findings to the class.

F ■ 55

ACTIVITY DOING

EXTINCT SPECIES

Skills: Applying concepts, making comparisons

This field activity involves a visit to a natural history museum and observation of extinct species. An alternative approach would be to simply assign a particular extinct organism to each student for library research. Students should report on the characteristics, time periods, and reasons for extinction for each organism they observe.

Integration: Use this Activity to integrate language arts into your lesson.

REINFORCEMENT/RETEACHING

Media and Technology

Focus students' attention on the importance of variation by using the video called Patterns of Evolution.

After viewing the video, lead an open discussion on the three main topics of the video.

• **Why were the gulls willing to raise the newly hatched young of another species?** (Because mating choice among the gulls is behavioral, not genetic.)

• **What is the fate of the cheetah?** (Answers will vary, but students should point out that the future for the cheetah is uncertain because of the lack of variation in its gene pool.)

• **Why were left and right spiraled snails in the same location able to evolve separately?** (Structural differences have prevented them from interbreeding.)

organisms seem closely related, others seem to be only distantly related. Scientists believe this is due to evolution.

CONTENT DEVELOPMENT

Point out that the molecular structure of protein molecules can also show relationships between organisms. Explain that because amino acids are used to build and repair body parts, the more similar the structure of the protein molecules of different organisms, the closer the relationship between the organisms.

● ● ● ● **Integration** ● ● ● ●

Use the discussion of molecular evidence of change to integrate concepts of proteins into your lesson.

ENRICHMENT

▶ *Activity Book*

Students who have mastered the concepts in this section will be challenged by the chapter activity Interpreting the Geologic Time Scale.

BACKGROUND INFORMATION
CHARLES DARWIN

After deciding against careers in medicine and the clergy, 22-year-old Charles Darwin was invited to serve as a naturalist on the *Beagle*, a small English ship. The *Beagle's* mission was to extensively survey South America and the Pacific.

During the *Beagle's* voyage, which began in 1831, Darwin collected large numbers of specimens and made detailed observations of the areas through which he traveled. He also read *The Principles of Geology* by Charles Lyell, which proposed that the Earth is very old and that the forces that have produced changes on the Earth's surface in the past are the same forces that operate in the present.

Lyell's ideas led Darwin to think that living things might also undergo changes over long periods of time. Darwin's ideas about animal and plant evolution were supported by his observations of the changes within species that he noted as he traveled along the South American coast and the Galapagos Islands.

Figure 2–9 *These strange-looking animals are not purely a product of the artist's imagination. They are based on fossils seven million years old. Like all living things, these animals evolved through modification of earlier life forms.*

the molecular clock can be used to determine how long ago the organisms branched off from that ancestor.

What do all these various types of evidence tell us? From fossils to molecular clocks, the answer is clear. **Living things have evolved through modification of earlier life forms. That is, living things have descended from a common ancestor.**

2–1 Section Review

1. Define evolution, using the term species in your definition.
2. Why are mutations called agents of change?
3. Describe the evidence that supports evolution.

Critical Thinking—*Applying Concepts*
4. Are birds' wings and butterflies' wings homologous structures?

2–1 (continued)

INDEPENDENT PRACTICE
Section Review 2–1

1. Evolution is a change in species over time.

2. A mutation is a change in a gene that will produce a change in the offspring of an organism. If the change is positive, the organism will be better adapted to its environment and produce more offspring. A sufficient number of mutations may result in a completely new species.

3. Possible answers include the following: Living things on Earth today are very different from the living things that existed in the past; the fossil record shows that living things have changed over time; species today that are very different can be shown to have developed from the same ancestor; homologous structures in living organisms indicate that they may have a common ancestor; similarities in the early stages of embryo development of different organisms also point to a common ancestor; similar DNA or protein structure in a living organism and an extinct organism can be used to show they both had a common ancestor.

4. Accept all logical responses. Students may need to do some research to find that birds' wings and butterflies' wings did not evolve from the same ancestral body parts.

REINFORCEMENT/RETEACHING

Review students' responses to the Section Review questions. Reteach any material that is still unclear, based on their responses.

2–2 Charles Darwin and Natural Selection

Guide for Reading

Focus on this question as you read.

▶ What is natural selection?

The Galapagos Islands rise out of the Pacific Ocean about 1000 kilometers from the west coast of South America. The islands received their name from the giant Galapagos tortoises that live there. The tortoises' long necks, wrinkled skin, and mud-caked shells make them look like prehistoric creatures. Sharing the islands with the tortoises are many other animals, including penguins, long-necked diving birds called cormorants, and large, crested lizards called iguanas.

The most striking thing about the animals of the Galapagos is the way in which they differ from related species on the mainland of South America. For example, the iguanas on the Galapagos have extra-large claws that allow them to keep their grip on slippery rocks, where they feed on seaweed. On the mainland, iguanas have smaller claws. Smaller claws allow the mainland iguanas to climb trees, where they feed on leaves.

In 1831, a young British student named Charles Darwin set sail for a five-year voyage on a ship called the *Beagle*. Serving as the ship's naturalist, Darwin studied animals and plants at every stop the ship made. When Darwin arrived at the Galapagos, he soon noticed many of the differences between island and mainland creatures. As he compared the animals on the mainland to those on the islands, he realized something special. It appeared that

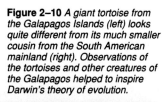

Figure 2–10 *A giant tortoise from the Galapagos Islands (left) looks quite different from its much smaller cousin from the South American mainland (right). Observations of the tortoises and other creatures of the Galapagos helped to inspire Darwin's theory of evolution.*

CLOSURE

▶ *Review and Reinforcement Guide*
Have students complete Section 2–1 in the *Review and Reinforcement Guide.*

TEACHING STRATEGY 2–2

FOCUS/MOTIVATION

Use a large world map to show students the route taken by the *Beagle* during Charles Darwin's voyage. Explain the purpose of the voyage.

CONTENT DEVELOPMENT

Emphasize that Darwin had already read other theories of evolution, including Lamarck's. As Darwin observed and examined the plant and animal life from different areas on his voyage, he saw many examples of change occurring within a species. Stress that Darwin noticed that these changes seemed to enable the organisms to survive well in their environment.

2–2 Charles Darwin and Natural Selection

MULTICULTURAL OPPORTUNITY 2–2

Have your students research what things are affecting the balance of nature in the rain forest. How does the elimination of one species affect other species? For example, what is the effect of the elimination of a predator species on the species that was formerly the prey?

ESL STRATEGY 2–2

Make sure that students understand the meaning and usage of the adjectives *fit, fitter,* and *fittest.* Help them form sentences using the three words. You may also want to discuss the verb *to fit* and the noun *fit,* and have students use these words in sentences.

Ask students to rearrange the four words below to form a sentence explaining Darwin's concept of natural selection.

the—selects—Nature—fittest

Challenge students to invent new titles for Darwin's book *On the Origin of Species.* Titles should retain the original meaning or reflect the main ideas of Darwin's work.

● ● ● ● **Integration** ● ● ● ●

Use the discussion of the Galapagos Islands to integrate concepts of geography into your lesson.

CONTENT DEVELOPMENT

Explain that Darwin spent over 20 years collecting more evidence to support his theory of evolution. Another British scientist, Alfred Russel Wallace, had developed ideas similar to those of Darwin through his work in Malaya. In 1858, both scientists published papers on the theory of evolution. A year later, Darwin published a book called *On the Origin of Species.* In this book Darwin presented an entirely new concept that he called natural selection.

● ● ● ● **Integration** ● ● ● ●

Use the discussion of Charles Darwin to integrate concepts of social studies into your lesson.

LAMARCK AND DARWIN

According to Lamarck's theory, variation is directed. Organisms deliberately change in response to the conditions of the environment and so become better adapted. In Darwin's theory, variation is basically undirected, or random. Organisms are born different from one another. By chance some variations are beneficial and others are not.

Darwin believed that there was some inheritance of acquired characteristics. He felt, however, that natural selection was a much more powerful force in the evolution of organisms.

HISTORICAL NOTE

ALFRED RUSSEL WALLACE

Just as Darwin was ready to publish his work, another English naturalist, Alfred Russel Wallace, came up with an idea that was virtually identical to Darwin's. Darwin and Wallace presented their papers at a scientific meeting. Both Wallace and the world at large agreed that Darwin deserved the lion's share of the credit for proposing the theory of evolution by natural selection because his work was more thorough.

2–2 (continued)

CONTENT DEVELOPMENT

Stress to students that all of the suggestions for the meaning of "survival of the fittest" may be correct, depending on environmental conditions. Explain that Darwin believed that only the organisms that were strong enough, fast enough, and/or smart enough to obtain food, water, and a place to live could survive. Point out to students that the process by which only the best-adapted members of a species survive is referred to as "survival of the fittest."

Explain to students that what Darwin meant by natural selection was that nature "selects" the fittest organisms.

Figure 2–11 *A green iguana's long toes, sharp claws, and green scales make it well suited for life in a tropical rain forest (left). Although they evolved from lizards similar to the green iguana, marine iguanas have webbed toes, thick claws, and brownish-gray scales. Their webbed toes and rounded snouts are adaptations for swimming (right).*

Activity Bank

Where Are They?, p.106

each animal was perfectly adapted to survival in its particular environment.

Darwin took many notes and collected many specimens. For the next 20 years he tried to find an underlying theory that could explain his observations. In 1858, Darwin and another British biologist, Alfred Wallace, presented independently a new and exciting concept—the theory of evolution.

This theory was discussed by Darwin in a book entitled *On the Origin of Species*. In this book, Darwin presented an entirely new idea—the concept of **natural selection.** Darwin used this concept to explain how evolution occurs, or the mechanics of evolution. **Natural selection is the survival and reproduction of those organisms best adapted to their surroundings.** To better understand how natural selection works, you must first learn about the role of overproduction in nature.

Overproduction and Natural Selection

Biologists have long known that many species seem to produce more offspring than can be supported by the environment. Every year, for example, dandelions grow seeds with sails that form into a white puff on the stem. The wind blows the seeds through the air. Most seeds land in a place where conditions are unfavorable for new dandelion growth. Only a few seeds land in a place with the right soil, light, and water conditions. These seeds grow into new dandelion plants. Through overproduction, nature assures that at least some seeds will survive to continue the species.

● ● ● ● **Integration** ● ● ● ●

Use the discussion of dandelions to integrate concepts of seed dispersal into your lesson.

FOCUS/MOTIVATION

Divide the class into teams of four to six students. Tell the teams to choose an edible animal such as a chicken, turkey, or beef cattle, or an edible plant such as potato, corn, or watermelon. Have the teams list new desirable characteristics of

the living thing that would make them more useful to people.

CONTENT DEVELOPMENT

Stress that many living things produce more offspring than can possibly be supported by an area's resources and that these offspring must vie with other species for food, water, and shelter. Explain that such overproduction increases the chances that at least some of an organism's offspring will survive the competition

Quite often, overproduction of offspring results in competition for food or shelter among the different members of a species. In the case of tadpoles, which hatch from frog eggs, competition can be fierce. The food supply in a pond often is not large enough for every tadpole to survive. Only those strong enough to obtain food and fast enough to avoid enemies will live. These animals will eventually reproduce. The others will die before producing offspring.

The process in which only the best-adapted members of a species survive is sometimes called survival of the fittest. In a sense, the fittest animals are selected, or chosen, by their surroundings to survive. This is basically what Darwin meant by natural selection—nature selects the fittest.

Variation and Natural Selection

Although the members of a species are enough alike to mate, normally no two are exactly the same. In other words, even members of the same species have small variations. For example, some polar bears have thicker coats of fur than others do. This thicker fur gives them more protection against the cold. Such polar bears are fitter, and thus more likely to survive and pass on the characteristic. In this case, a variation in a species will cause some members to survive and reproduce. Over time, the variation will become the norm as those members of the species with the variation survive in greater numbers than do those members without the variation. Can you think of another example of variation in a species? ②

Figure 2–12 *Living things, such as scorpions and dandelions, produce many more offspring than survive. What is the name of the process that determines which individuals survive and reproduce? How does it work?* ①

Activity Bank

Variety Is the Spice of Life, p. 108

ACTIVITY DISCOVERING

Survival of the Fittest

Scatter a box of red and green toothpicks in a grassy area. Then have a friend pick up in 10 minutes as many toothpicks as he or she can.

Was one color toothpick collected more than the other color? If so, explain why.

■ How can a variation such as color affect the process of natural selection?

F ■ 59

60 ■ F

Figure 2–13 *Although these flamingoes may look alike, there are some variations among them. The flamingoes with the characteristics best suited for their environment survive and reproduce, passing on their characteristics to the next generation.*

Figure 2–14 *Natural selection may produce some amazing results. The markings and behavior of this caterpillar trick hungry birds into thinking that it is a bird-eating snake. The color and irregular shape of the horned frog make it hard to see among the leaves on the forest floor.*

In the same way, members of the same plant species may show minor variations in the length and thickness of roots. Plants with deeper root systems can reach under ground more easily and thus will have a better chance for survival than do plants with shorter root systems. In this example, the plants with deeper roots are more likely to be "selected" by nature and to pass on their traits to generations that follow. The plants with shallow root systems are not quite as fortunate. These plants would have a better chance for survival in an area in which most water was close to the surface. As you can see, variations among members of a species are another reason natural selection can lead to changes in living things over time.

Minor variations in a species are common. Sometimes, however, mutations can cause a change in an organism's characteristics that is far from minor. For example, in White Sands National Monument in New Mexico, the sand dunes are white. White mice live on these dunes. The light color of the mice is a result of a helpful mutation. Because white mice blend in better with their environment than darker mice do, they are less likely to be eaten by predators. If a mutation occurred that darkened some of the mice, these darker mice would not be able to blend in with their surroundings. The mutation that caused the darker color would be considered a harmful mutation. In this case, natural selection would "weed out" the mice with the harmful mutation. As generations of mice reproduced, the darker mice, along with their harmful mutation, would be eliminated from the species. The white mice with the helpful mutation would survive and multiply.

2–2 (continued)

CONTENT DEVELOPMENT

Remind students that although most members of a species are much alike, there are still some variations. Explain that some of these variations can enable one member of a species to survive better than another member. Explain that the natural design of the environment selects the type or kind of organism that can live there.

• **What do you predict would happen if a plant such as a cactus that needs very little water were planted in a swamp?** (Accept all logical answers.)

• **What do you predict would happen if a furry animal such as a moose were placed in a desert?** (Accept all logical answers.)

Media and Technology

Students can observe how changes in an ecosystem can affect the organisms within that ecosystem when they investigate ecological changes that may have caused the ducks to disappear in the Interactive Videodisc/CD ROM called Paul ParkRanger and the Mystery of the Disappearing Ducks. Using the media, they will explore Paul's cabin, which is full of clues to the disappearance, as well as interview experts and local people, all of whom have a different opinion on the cause of the ducks' disappearance. When students complete the Videodisc, they should hypothesize why the ducks disappeared and discuss how changes in an ecosystem can have far-ranging effects on living things.

● ● ● ● **Integration** ● ● ● ●

Use the discussion of the British peppered moth to integrate concepts of social studies into your lesson.

INDEPENDENT PRACTICE

▶ *Activity Book*

Students who need practice with the concepts of this section should be provided with the chapter activity Modifying the Manatee.

GUIDED PRACTICE

▶ *Laboratory Manual*
Skills Development

Skills: Applying concepts, making observations, interpreting charts

At this point you may want to have students complete the Chapter 2 Laboratory Investigation in the *Laboratory Manual* called Variation and Natural Selection. In the investigation, students will observe variations in plants and develop a model illustrating natural selection.

Mirrors of Change

In some ways living things become a mirror of the changes in their surroundings. The British peppered moth is a recent example of this phenomenon. In the 1850s, most of the peppered moths near Manchester, England, were gray in color. Only a few black moths existed. Because the gray moths were almost the same color as the tree trunks on which they lived, they were nearly invisible to the birds that hunted them for food. Most of the black moths, however, were spotted by the birds and eaten. The species as a whole survived because of the gray moths. Then changes in environmental conditions had a drastic effect on the moths that lived in the area.

As more factories were built in the area, soot from the chimneys blackened the tree trunks. The gray moths could now be seen against the tree trunks. The few surviving black moths, however, now blended in with the tree trunks. As a result, they survived. These moths produced more black offspring. In time, practically all peppered moths were black. Again, the species as a whole survived.

The tale of the peppered moths illustrates how natural selection was able to turn an unusual trait into a common one in a relatively short period of time. As pollution controls in England become tighter in the years to come, how might the peppered moth species be affected? ❶

Figure 2–15 *After the start of the Industrial Revolution, the light gray bark of trees was darkened by the soot from factories. In each photograph, which peppered moth would most likely be noticed by a hungry bird? How did pollution affect the way that natural selection acted upon peppered moths?* ❷

2–2 Section Review

1. Why is natural selection another way of saying survival of the fittest?
2. What is the relationship between overproduction of offspring and natural selection?
3. Describe how living things can become a mirror of the changes in their environment.

Connection—*Ecology*
4. There is evidence that Earth's climate is getting warmer. This phenomenon is called global warming. How might global warming affect the evolution of living things?

BACKGROUND INFORMATION
DARWIN AND RELIGION

Despite efforts by antievolutionists to paint Darwin as an atheist, he was a very religious man. He had a degree in theology from Cambridge and even thought about spending his life as a country preacher. He remained a devout Christian all his life. He saw no conflict—and no less awe or majesty—in a God that ruled through natural laws rather than through supernatural means.

sulting adaptations may eventually result in a new species.

4. Individual organisms that are better adapted to a warmer climate will be favored by natural selection. Some species—those that can only live in a cooler climate—may be unable to adapt and will become extinct.

CONTENT DEVELOPMENT

◉ **Media and Technology**

To demonstrate how animals adapt to various ecosystems have students use the videodisc called Insects: Little Giants of the Earth.

After students have used the videodisc, have them write brief essays describing the ways in which insects adapt to various ecosystems.

INDEPENDENT PRACTICE
Section Review 2–2

1. In both survival of the fittest and natural selection, organisms best adapted to their environment survive and reproduce.
2. When overproduction of a species occurs, natural selection brings about the survival of only the fittest individuals.
3. If the environment changes in a way that threatens a species, individuals better adapted to the new environment may survive and increase in number. The re-

REINFORCEMENT/RETEACHING

Review students' responses to the Section Review questions. Reteach any material that is still unclear, based on their responses.

CLOSURE

▶ *Review and Reinforcement Guide*
Have students complete Section 2–2 in the *Review and Reinforcement Guide*.

2-3 The Development of a New Species

MULTICULTURAL OPPORTUNITY 2-3

For an interesting study of new species, study Australia and New Zealand. These two areas of the world have many species of animals that are unique. Also unique to Australia are the aboriginal peoples. Have your students investigate the close relationship that these people have with their environment.

ESL STRATEGY 2-3

Point out the similarity in endings of the three underlined words in the following statement. Migration and isolation lead to speciation. Then ask students to write the definition of each of the three words.

Mention that another term ending in the suffix -ation is adaptive radiation. Ask students if they believe that the term means that organisms can adapt to radiation. Have them support their answers with logical arguments.

Guide for Reading

Focus on these questions as you read.

▶ Under what conditions do new species evolve?

▶ What is adaptive radiation?

2-3 The Development of a New Species

You have seen how, through natural selection, the fittest organisms survive and reproduce. Natural selection explains how variations in a species can lead to changes in that species. But how does an entirely new species evolve? Recall that a species is a group of organisms that share similar characteristics and that can interbreed and produce fertile offspring. In order to understand how a new species forms, you have to know a bit more about competition among organisms.

Niche

Every type of living thing has certain needs that must be filled in order for that living thing to survive. For example, organisms need food, shelter, and water. The combination of an organism's needs (which must be supplied by the environment), its habitat (where it lives), and the role it plays in its habitat (how it affects and is affected by the living and nonliving things around it) is called the organism's **niche**. A lion's niche, for example, is found in 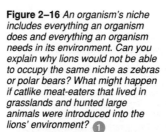 the plains of Africa. A lion would not survive outside its niche—for example, in the cold arctic climate. A polar bear's niche, on the other hand, is found in the cold arctic. A polar bear would not survive on the African plains.

Figure 2–16 *An organism's niche includes everything an organism does and everything an organism needs in its environment. Can you explain why lions would not be able to occupy the same niche as zebras or polar bears? What might happen if catlike meat-eaters that lived in grasslands and hunted large animals were introduced into the lions' environment?* ❶

TEACHING STRATEGY 2-3

FOCUS/MOTIVATION

In human cities thousands of people survive near one another. They have different jobs, they shop in different stores, and they live in different places. Animals and plants do much the same thing. The combination of an organism's "profession" and the place in which it lives is called its niche. If two species occupy the same niche in the same location at the same time, they will compete with each other for food and space. One of the species will not survive. No two species can occupy the same niche in the same location for a long period of time.

CONTENT DEVELOPMENT

Encourage students to think about what it means to be isolated. You might use the familiar tale of being shipwrecked on a deserted island.

● ● ● ● **Integration** ● ● ● ●

Use the discussion of niches to integrate concepts of ecology into your lesson.

GUIDED PRACTICE

Skills Development

Skill: Applying concepts

Ask students to write brief autobiographies. They should include where they live, something about their school life, and information about their activities and hobbies; in other words, the point is to establish their "niche."

FOCUS/MOTIVATION

Use the student autobiographies to introduce the concept of biological niches. Emphasize that no two autobiographies are exactly alike. Neither can two species occupy the same place and function in a biological system and be successful. Bring up the ideas of competition and struggle

One general rule of biology is that when two organisms occupy the same niche, they strongly compete with each other for food, shelter, and water. If two organisms occupy different niches, they do not strongly compete with each other. If two species of birds, for example, live in the same tree, but one species lives in the upper branches and feeds on insects found on those branches and the other species lives on the lower branches and feeds on different insects on those branches, then the two species occupy different niches. They live in the same tree, but they are not in competition for the same niche. As you continue to read about niches, keep in mind that a niche can be very small and specific.

When two species occupy the same niche, one species will be successful and survive and the other species will be less successful and possibly become extinct. When two species occupy different niches, they both have a better chance for survival. **In general, new species evolve when there are empty niches that can be filled or when a species moves into a niche it did not previously occupy.**

Migration and Isolation

There are two common ways in which organisms may move into a new or empty niche. One way is through migration. In simple terms, migration means moving from the place in which an organism lives to a new home. During Earth's long history, many organisms have migrated to new areas. Once in a new area, the organism may move into a new niche. If the niche is empty, there will be no competition. If the niche is not empty, competition will arise. Either the new organism or the organism that originally lived in that niche will die off.

The other way an organism may occupy a new niche is through isolation. Isolation occurs when some members of a species suddenly become cut off from the rest of that species. Isolation may be due to barriers that form over time. For example, when a mountain range rises, the organisms living on each side of the range become isolated from one another. (Sure that takes a long time, but evolution can be a very long process.) As the members of a species become isolated from one another, they may begin

Warmblooded Dinosaurs?

For many years, scientists believed that dinosaurs were coldblooded, as are modern reptiles. Now, however, there is much evidence that at least some dinosaur species were warmblooded. Using reference materials, write a report on warmblooded dinosaurs, providing evidence for and against the theory. At the end of your report, give your opinion. You must back up your opinion with solid scientific evidence.

F ■ 63

ACTIVITY WRITING

WARMBLOODED DINOSAURS?

Explain to students that animals can be roughly classified into two groups (coldblooded ectotherms or warmblooded endotherms) based on how they generate and control their body heat.

Students will discover that the question posed in the Writing Activity is hotly debated among scientists. Those supporting the coldblooded theory point out that it is easier to explain the extinction of dinosaurs if it is assumed they were ectotherms. In contrast, scientists supporting the warmblooded theory use the fact that certain features of dinosaur bones make them quite similar to mammalian bones.

Integration: Use this Activity to integrate language arts into your lesson.

for existence. Emphasize that the result is frequently extinction of a species and sometimes creation of a different species.

CONTENT DEVELOPMENT
Media and Technology

Students will gain a far better understanding of the concept of niche when they explore the relationships among various organisms and their habitats in the Interactive Videodisc/CD ROM called Amazonia. Exploring the differing layers of a tropical rain forest, students will develop a clear understanding of how various organisms are well adapted to a particular, and often rather exclusive, niche within the tropical rain forest. When they complete the videodisc, have them hypothesize what would happen to the organisms living in specific niches when large plots of rain forest are cleared for farming.

Figure 2–17 *In Australia, pouched mammals evolved to fill niches that are occupied by other types of mammals elsewhere in the world. The kangaroo, surprisingly, is the Australian equivalent of an antelope—a large, fast-moving plant-eater that lives in large groups. The cat-sized cuscus lives in trees, eats fruit and small animals, has a grasping tail, and is quite curious. What familiar animal occupies a niche similar to that of the cuscus?* ❶

to fill different niches, particularly if the geography and climate in the area change as well.

In Chapter 1 you read about a very dramatic example of isolation. Hundreds of millions of years ago, all of Earth's landmasses were combined in the supercontinent called Pangaea. Over time, Pangaea split into the continents that exist on Earth today. As ❶ the continents separated, species became isolated from one another and began to occupy new niches. Perhaps the most striking illustration of this process of isolation is the continent of Australia. The organisms living in Australia have been isolated from all other organisms on Earth for millions of years. As a consequence, organisms in Australia have evolved in different ways from organisms in the rest of the world. And that is why many of the living things in Australia are so different from living things found almost everywhere else on Earth.

Speciation: Occupying New Niches

Scientists use the term speciation to describe the development of a new species. Perhaps the best way to explain how migration, empty niches, and isolation can lead to the evolution of a new species is through an example. Imagine that a species of birds lives in a particular area. All the birds eat the tiny seeds of a shrub that grows in that area. Now imagine that through barriers or perhaps through the movement of continents, the birds become separated into three different areas. In the first area, the shrubs with the seeds the birds eat still thrive. In the

---

Because of genetic drift and other factors, some characteristics may have little or no effect on an individual's fitness. There is no indication, for example, that hair color in humans is or ever was important to survival. Yet many different shades of hair color are found in the human population.

On the other hand, some adaptations decrease an organism's fitness in some respects and increase it in others. For example, a peacock's long colorful tail feathers require much energy to grow, make the peacock easy for predators to spot, and interfere with the peacock's ability to fly. The elaborate plumage, however, increases a peacock's fitness by making him more attractive to females. Although a brightly colored peacock is much more vulnerable to predators, he will reproduce and pass on his genes more often than a drab, if longer-lived, peacock.

2–3 (continued)

CONTENT DEVELOPMENT

Remember that a species is a group of organisms that can breed with one another and produce fertile offspring in a natural environment. This means that individuals in the same species share a common gene pool. Individuals in different species have different gene pools.

● ● ● ● **Integration** ● ● ● ●

Use the discussion of Pangaea to integrate concepts of continental drift into your lesson.

GUIDED PRACTICE

Skills Development

Skill: Making observations

Begin this activity by briefly discussing the theory of continental drift. Explain that many scientists believe that at one time the Earth was one giant landmass that split apart to form the continents. It is believed that large currents of molten

rock under the surface of the Earth are responsible for this movement. To demonstrate continental drift, fill a large, deep casserole dish with water. Place the dish on a heat source and add several drops of vegetable coloring to the water.

• **What happens to the vegetable coloring?** (It follows a convection current formed in the water as it is heated.)

Place several balsa wood blocks in the center of the heated water.

• **What happens to the blocks of wood?** (They move away from one another, following the convection currents.)

• **How is this activity related to the theory of the movement of the continents?** (Large convection currents can move continents away from one another just as these small currents moved the pieces of balsa wood.)

CONTENT DEVELOPMENT

Scientists have learned that new species usually form only when populations are

second area, the shrubs are no longer common. Instead, different shrubs with larger seeds grow. In the third area, very few seed-bearing shrubs exist. But shrubs with berries grow extremely well.

Now assume that before the isolation, the beaks of the birds were well adapted to picking up and breaking open the small seeds of the shrubs. The birds that have been isolated in the area with the same shrubs still occupy the same niche. Thus they will probably not change much over time. They will be successful by staying in the niche they occupy.

As you have just read, the shrubs in the second area have larger seeds. The birds there have a difficult time picking up and breaking open the seeds. However, one day a mutation or variation in the birds produces a bird with a larger, stronger beak. This bird will be more successful at eating the larger seeds. Over time, the birds with the small beaks will die off as nature selects the fittest birds for survival. Only those birds with the larger, stronger beaks will survive and reproduce.

In the third area, a mutation or variation produces an offspring with a beak well adapted to picking and eating berries. Again, the birds with the small beaks will die off as nature selects the birds with the better adapted beaks—the fitter birds. What do you think might happen to birds born with this adaptation who live in the first area where the shrubs with tiny seeds still grow? ❷

As you can see, over time the beaks of the birds will have changed. The birds in each area will be better adapted to their environment because of the

Figure 2–18 *Natural selection is the most important force behind evolution. However, some changes may occur simply by chance. Natural selection provides an advantage to rhinoceroses with horns. But the African rhinoceros (top) is not more fit than the Indian rhinoceros (bottom) because it has two horns rather than one. During the course of evolution, the two types of rhinoceroses developed different numbers of horns purely by chance.*

result in organisms that breed only with individuals that are most similar to themselves.

Once reproductive isolation occurs, natural selection usually increases the differences between the separated populations. As the populations become better adapted to different environments, their separate gene pools gradually become more dissimilar. If the populations remain separated for a long time, their gene pools eventually become so different that their reproductive isolation becomes permanent. When this occurs the groups of organisms are no longer separate populations. They have become separate species.

isolated, or separated. This separation of species prevents the populations from interbreeding to produce fertile offspring.

● ● ● ● **Integration** ● ● ● ●

Use the discussion of the mutation and variation of birds to integrate the concept of adaptations into your lesson.

ENRICHMENT

The separation of populations that prevents interbreeding is called reproductive isolation. If two populations are not reproductively isolated, their gene pools will blend with each other. No new species will be formed. Reproductive isolation is the agent for the formation of new species.

Reproductive isolation may occur in a variety of ways. Geographic barriers such as rivers, mountains, and even roads may separate populations and prevent them from interbreeding. Differences in courtship behavior or fertile periods may

LOVELOCK CHALLENGES NATURAL SELECTION

Dr. James E. Lovelock has developed a hypothesis that has changed the way many scientists view the Earth. Named after the Greek goddess of the Earth, the Gaia hypothesis states that the Earth is like a living system. Not only does the environment of the Earth affect living things, but living things also affect the Earth's environment. Living things help keep the Earth's environment stable over time. This stability, in turn, helps living things survive.

Lovelock points out several examples of the Earth's homeostasis. The sun's energy output has increased more than 30 percent during the past 3.5 billion years, yet the average temperature on the Earth has changed very little. The salinity of ocean water remains fairly constant at 3.5 percent despite the billions of kilograms of salt that are carried into the ocean by rainwater every year. The oxygen level of the atmosphere remains constant at about 21 percent. If it rose to 25 percent, fires would burn uncontrollably; if it dropped slightly, many organisms would die.

Most evolutionary biologists do not agree with Lovelock because the Gaia hypothesis runs counter to the laws of natural selection. Geochemists argue that living organisms contribute very little to global chemistry. What do you think?

2–3 (continued)

CONTENT DEVELOPMENT

The process of adaptive radiation is also known as divergent evolution. In adaptive radiation, a number of different species diverge, or move away, from a common ancestral form, much as the spokes of a bicycle wheel radiate from the hub. During a period of adaptive radiation, organisms evolve a variety of characteristics that enable them to survive in different niches.

Throughout the history of life on Earth, adaptive radiations have occurred many times and in many places. Adaptive radiation occurred on the Hawaiian islands among a group of birds called Hawaiian

honeycreepers. The dinosaurs experienced an adaptive radiation in their day, only to eventually become extinct. The mammals alive today were produced by another wave of adaptive radiation.

INDEPENDENT PRACTICE

▶ *Activity Book*

Students who need practice with the concepts of this section should be provided with the chapter activity Adaptive Radiation.

changes in the beaks. Furthermore, although this example has concentrated on changes in beaks, you can probably assume that other structural changes will also have occurred as each bird species became better adapted to its environment. Over time, the birds will have changed so much that they can no longer interbreed and produce fertile offspring. In

Figure 2–19 *Adaptive radiation is the process by which many different species develop from a common ancestor. As you can see, some of the descendants of the cotylosaur do not resemble it at all!*

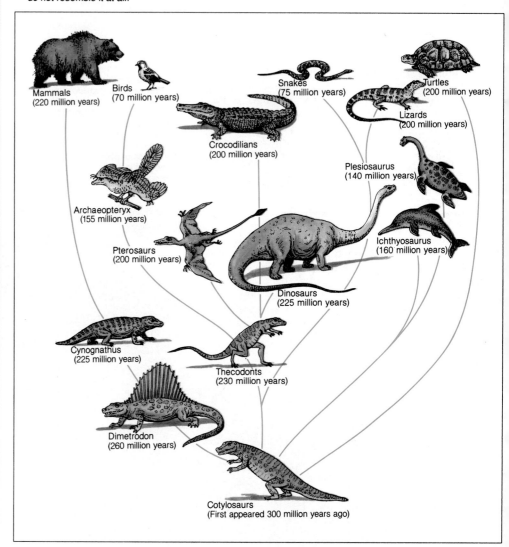

INDEPENDENT PRACTICE

Section Review 2–3

1. They compete with each other for food, shelter, and water. One species may become extinct.

2. Some members of a species either move away (migration) or become separated (isolation) from the original population. If the new environment is different, the separated population may adapt enough over time that it becomes a new species.

other words, several new species of birds will have evolved from the original species.

Although this example has been imaginary, it is quite similar to the situation Darwin encountered in the Galapagos. On the islands of the Galapagos, Darwin found many species of a type of bird called a finch. On each island, the finches were slightly different and had formed a new species. Darwin's finches—just as the birds in our example—had evolved, and those better adapted to each island had survived.

Speciation and Adaptive Radiation

The process in which one species evolves into several species, each of which fills a different niche, is called adaptive radiation. In **adaptive radiation**, organisms of a species "radiate," or move away from, other organisms in that species and occupy new niches. Over time, these organisms may evolve into entirely new species. Keep in mind that in adaptive radiation, the new species all share a common ancestor. You might say they adapted to their new environment as they radiated away from the area where their common ancestor lived. The homologous structures you have read about earlier are evidence of such adaptive radiation in which similar body parts of related organisms evolved to perform different functions.

2–3 Section Review

1. What happens when two species occupy the same niche?
2. What is the relationship between migration, isolation, and speciation?
3. Define adaptive radiation. Use the term niche in your definition.

Connection—*Ecology*

4. As people have moved from place to place, they have often brought plants and animals with them. How might the introduction of a new species of plant or animal in an area have disastrous effects on the organisms already living in that area?

Living Dinosaurs?

Many biologists assert that the dinosaurs are not extinct, but rather are alive and well. These biologists think that birds are actually modern-day dinosaurs.

Current theory indicates that birds evolved from the most famous of the dinosaurs, *Tyrannosaurus rex.* Using reference materials, find out why birds may be related to *Tyrannosaurus rex.* (*Hint:* You will concentrate on jawbones.)

On posterboard, illustrate those similarities and differences you have discovered between dinosaurs and modern birds.

F ■ 67

ACTIVITY
DOING

LIVING DINOSAURS?

Skills: Applying concepts, making comparisons, relating cause and effect

Ask many paleontologists what a bird is and they'll reply with a grin, "a hot-blooded dinosaur with feathers." The first fossil ever found of an early birdlike animal is called *Archaeopteryx* and dates from late in the Jurassic Period. Its skeleton looks much like a small running dinosaur.

Unlike modern birds, *Archaeopteryx* had teeth in its beak. It also had toes and claws on its wings. *Archaeopteryx* would be classified as a dinosaur except for one important feature: It had well-developed feathers covering its entire body. *Archaeopteryx* may have descended from the same dinosaur line that eventually gave rise to *Apatosaurus* and its kin.

Integration: Use this Activity to integrate fine arts into your lesson.

BACKGROUND INFORMATION

FRUIT FLIES REPLACE FINCHES?

Darwin's finches may finally get a rest! Evolutionary biologists have a new case to use in studying divergence of species. Native Hawaiian fruit flies have been found singing an assortment of courtship songs. The more than 500 different species make sounds that the other species on the island cannot recognize. Some Hawaiian *Drosophila* sound more like cicadas than flies. Others make a cricketlike noise. Still others make a sound like that of a North American fruit fly, but they create the sound by vibrating the abdomen instead of the wings.

3. Adaptive radiation is the process by which one species evolves into several new species, each filling a different niche.
4. Accept all logical answers. The introduced organism might not have any natural enemies in the new environment. It could move into and take over a niche occupied by a native organism.

REINFORCEMENT/RETEACHING

Review students' responses to the Section Review questions. Reteach any material that is still unclear, based on their responses.

CLOSURE

▶ *Review and Reinforcement Guide*
Have students complete Section 2–3 in the *Review and Reinforcement Guide.*

All of the Galapagos Islands' finch species studied by Darwin evolved from a single ancestral species. Yet each of the species exhibits body structures and behaviors that enable it to live in a different niche.

For example, each species shows adaptations that allow it to feed differently. Some of the finch species eat small seeds, whereas others crack open much larger seeds or seeds with thicker shells. Some species pick ticks—small insectlike animals—off the islands' tortoises and iguanas. One finch species uses twigs or cactus spines to remove insects from inside dead wood. And some finches, often called vampire finches, drink the blood of large sea birds after pecking them at the base of their tail!

Have students study the chart in the Problem Solving feature and then use it to answer the questions.

1. There are 12 species pictured on the diagram.

2. The primary difference is their habitat, either trees or the ground. The tree-finches eat insects or are vegetarians. The ground-finches are cactus-eaters.

3. *C. crassirostris.*

4. *G. conirostris* and *G. scandens.*

5. It may eat a different type of food, probably seeds.

6. The various species of finches on the Galapagos Islands show how geographic and behavioral barriers and reproductive isolation eventually lead to the formation of new species.

PROBLEM Solving

Darwin's Finches

While in the Galapagos Islands, Charles Darwin conducted detailed studies of the finches that lived there. The illustration below shows the possible evolution of some of the finch species that Darwin studied. The scientific name and beak outline are shown for each species.

Interpreting Evidence

1. How many species of ground finches are present in the Galapagos Islands?
2. What is the primary difference between the main types of finches?
3. Each species has a common name as well as a scientific name. What is the scientific name for the vegetarian tree finch?
4. Which ground finches are cactus-eating finches?
5. Suggest a reason why the sharp beak of the *G. difficilis* differs from the beaks of other ground finches.
6. How does the chart demonstrate evolution?

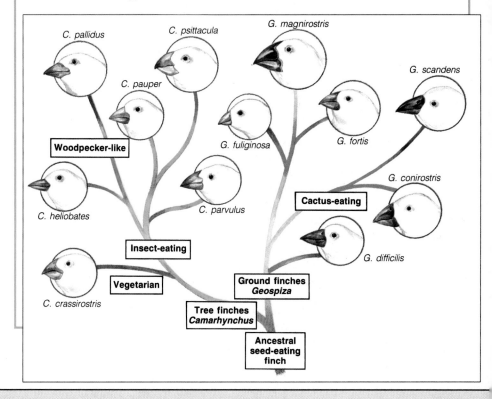

TEACHING STRATEGY 2–4

FOCUS/MOTIVATION

Inflate and tie two balloons with string. Tie the balloons to the ends of a meterstick. Place a fulcrum near the middle and adjust the meterstick to make it balance. Discuss the demonstration.

• **What do you observe?** (Accept all logical answers. The balloons are balanced.)

Point out that when things are balanced, they are said to be in equilibrium.

Explain that things in equilibrium are not always exactly balanced. Move your hand to make the meterstick teeter back and forth.

• **How does the equilibrium of the system change?** (Accept all logical answers. When one side goes down, the other side goes up and then they reverse.)

Have a student puncture one balloon with a pin.

• **What happened to the equilibrium?** (It changed rapidly.)

CONTENT DEVELOPMENT

Emphasize that Darwin believed that organisms evolved gradually over time. This is undoubtedly the case in most instances. This theory has become known as gradualism. Punctuated equilibrium is a most interesting and still controversial theory. Allow students the opportunity to discuss and debate this issue.

2-4 Punctuated Equilibrium

In Chapter 1 you read about the dinosaurs. Dinosaurs dominated the Earth for more than 150 million years. Then, about 65 million years ago, all living species of dinosaurs became extinct. (Around the same time, about 95 percent of all other living things also became extinct.) Scientists call such extinctions mass extinctions.

Mass extinctions seem rather harsh. After all, suddenly a great many species disappear, never to be seen again. In evolutionary terms, however, mass extinctions play an important role in the development of new species. After a mass extinction, a wide variety of previously occupied niches become available to those species that still exist. And as you might expect, many adaptive radiations occur after a mass extinction, as species move into new niches. In fact, after the mass extinction of the dinosaurs, a type of living thing that had existed for more than 50 million years in the shadow of the dinosaurs underwent a great adaptive radiation. This type of living thing began to fill unoccupied niches throughout the world. Do you know what this type of living thing is? You are correct if you said mammal. You are a mammal. So is a dog, a lion, a whale, and a skunk. Mammals are the dominant life form on Earth today primarily because the dinosaurs died off, leaving so many niches for mammals to fill.

In most cases, natural selection as described by Darwin is a long, slow process. Scientists do not doubt that natural selection occurs or that it leads to the evolution of living things. However, the fossil record shows very little evidence of gradual change. (Remember, most organisms do not leave fossils. So this lack of fossil evidence is to be expected.) But the fossil record does seem to indicate that some species may not change at all for long periods of time. This period of stability, or equilibrium, may continue for millions of years. Then suddenly, a great adaptive radiation may occur and a species may evolve into many new species, filling new niches. The equilibrium is broken, or punctuated.

In 1972, scientists Stephen Jay Gould and Niles Eldridge developed a theory called **punctuated equilibrium.** As you read its definition, keep in mind

Guide for Reading

Focus on this question as you read.

▶ What is punctuated equilibrium?

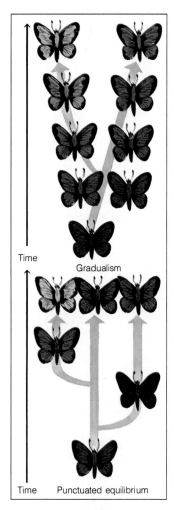

Time

Gradualism

Time   Punctuated equilibrium

Figure 2–20 *Organisms may change slowly and gradually over time. Or they may remain the same for a long time, then change quickly and abruptly.*

2-4 Punctuated Equilibrium

MULTICULTURAL OPPORTUNITY 2-4

Have your students brainstorm various kinds of changes that they observe. Then have them classify these as gradual changes (e.g., aging, the phases of the moon, tides) or sudden changes (e.g., changes in a person's mood, volcanic eruptions or earthquakes, changes in the weather). Make an analogy between gradual versus sudden changes and the slow process of evolution versus the sudden occurrence of punctuated equilibrium.

ESL STRATEGY 2-4

Explain the term *punctuated equilibrium* by using the ideas of "broken stability" or "interrupted balance."

Explain the expression *mutually exclusive.* Then ask students if scientists believe that both forms of evolution, punctuated equilibrium and gradual change, cannot occur at the same time. Are the two theories mutually exclusive?

GUIDED PRACTICE

Skills Development

Skill: Interpreting illustrations

Have students observe Figure 2–20 and read the caption.

• **What is being compared in the illustration?** (Two different theories that explain evolution, gradualism, and punctuated equilibrium.)

Point out that the top part of the illustration shows the butterflies changing color gradually over time.

• **How is this different from the color changes shown in the second part of the illustration?** (The color changes are much more abrupt.)

• **How does this illustrate the theory of punctuated equilibrium?** (The color of the butterflies remains stable for long periods of time. Then it makes great changes suddenly.)

REINFORCEMENT/RETEACHING

Have students make illustrations similar to those in Figure 2–20 using finches

instead of butterflies. The two parts of the illustration should compare Darwin's finches on the Galapagos Islands (punctuated equilibrium) with another group of birds that evolved according to the theory of gradualism. You may wish to have students do research to find a specific example of gradualism. Or they might invent an imaginary scenario.

BACKGROUND INFORMATION

THE "EVOLUTION" OF EVOLUTIONARY THEORIES

The work of both Hutton and Lyell was critical to Darwin's thinking about the nature and rate of change over time. In Darwin's day, creationist thinking insisted that neither the Earth nor the organisms on it had changed since their creation. Darwin, therefore, had to overcome a deep-seated belief in the fixity of nature. Lyell's work convinced him that the Earth had changed over time, setting him up to realize that organisms had changed, too.

Later on, however, Darwin's exclusive belief in gradualism led him to certain problems. Whether or not the punctuated equilibrium theory continues to gain acceptance, it is clear that rates of evolutionary change vary over time and among species.

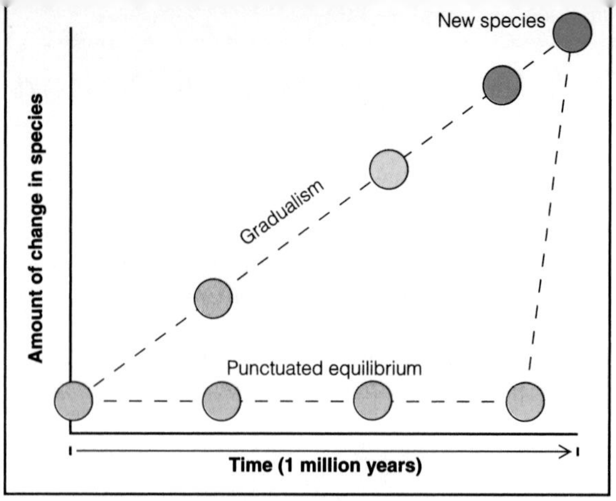

2–4 (continued)

FOCUS/MOTIVATION

Explain that a species arises every now and then that is particularly well adapted to its environment. When this is the case, the species may not change very much over time. One example of such a species is the horseshoe crab, *Limulus,* whose living members are nearly identical to ancestors that lived hundreds of millions of years ago. If possible, show a specimen of the horseshoe crab to students and explain how little it has changed over time.

Explain that organisms such as the horsehoe crab are relatively rare among both plants and animals. Sometimes such organisms are called "living fossils."

Figure 2–21 *Darwin thought that evolutionary change always occurs slowly and continually. This traditional view of the rate of evolution was challenged by the theory of punctuated equilibrium. In some cases, according to this theory, evolution may be at a standstill most of the time. These long periods of stability are occasionally interrupted by bursts of rapid change.*

Figure 2–22 *Natural disasters such as forest fires create new niches and may open up old ones. How have large-scale natural disasters shaped the evolution of life on Earth?* ①

• **Why is the term** *living fossils* **appropriate for the horseshoe crab?** (Accept all logical responses. Students may suggest that the living animal can provide information about life in ancient times in a way similar to information from the fossil record.)

CONTENT DEVELOPMENT

Point out that Darwin believed that species changed slowly and steadily over a long period of time. Tell students he re-

---

that in evolutionary terms, "short" can mean thousands of years. **According to punctuated equilibrium, there may be periods in Earth's history in which many adaptive radiations occur in a relatively short period of time.**

When does punctuated equilibrium seem to occur? As you might expect, it is most common when many niches are opened. This occurs during a mass extinction. It can also occur as a result of isolation, such as the isolation of organisms in Australia.

Today scientists believe that evolution can occur gradually—as described by Darwin—as well as fairly rapidly—as described by Gould and Eldridge. Neither theory disputes the other, and both seem to be valid. That is, punctuated equilibrium does not mean gradual change is incorrect. And Darwin's theory of gradual change does not mean that punctuated equilibrium is incorrect. Both forms of evolution seem to have occurred during Earth's 4.6-billion-year history.

2–4 Section Review

1. What is punctuated equilibrium? What is a mass extinction? How are they related?

Critical Thinking—*Making Inferences*
2. In what ways does the fossil record support punctuated equilibrium?

---

ferred to this type of evolution as gradualism.

Have students observe Figure 2–21. Tell students that scientists have found very few fossil records that support the theory of gradualism.

Point out that a second theory developed by Stephen Gould and Niles Eldridge states that a species may change very little or not at all for a long period of time and then have a sudden or rapid change resulting in a new species. Point

CONNECTIONS

Planetary Evolution ❶

While we tend to think of evolution as a biological event, the term is relevant to other sciences as well. Astronomers, for example, talk of the evolution of our solar system. The current theory on the evolution of our solar system is called the *nebular theory.*

According to the nebular theory, our solar system evolved from a huge cloud of dust and gas called a nebula. Shock waves, probably from the explosion of a nearby star, disrupted the dust and gas in the nebula. In reaction to the shock waves, the gases in the nebula began to contract inward, causing the nebula to shrink. As it shrank, the dust and gases began to spin around the center of the nebula. In time, the spinning nebula flattened into a huge disk almost 10 billion kilometers in diameter.

Near the center of the disk a new sun or *protosun* began to form. Gases and other matter surrounding the newly formed sun continued to spin. Some of the gas and matter began to clump together. Small clumps became larger and larger clumps. The largest clumps became *protoplanets,* or the beginnings of planets.

Over time the protoplanets near the sun became so hot that most of their gases boiled away, leaving behind a rocky core. Today these planets are known as Mercury, Venus, Earth, and Mars. Planets farther from the sun did not lose their gases because they did not receive as much heat. These planets became the gas giants. Today these gas giants are known as Jupiter, Saturn, Uranus, and Neptune.

Now here's a question for you to try to answer: How do you think Pluto, the ninth planet, formed? *Hint:* Pluto is a cold, barren world much different from the outer gas giants.

CONNECTIONS
PLANETARY EVOLUTION

Before students read the Connections feature, you may wish to provide some background information on the nebular theory, the belief that the solar system began as a huge cloud of dust and gas called a nebula.

The nebula in which the solar system formed would probably still be here today if a supernova in a nearby star had not disrupted the gases and dust of the nebula. The explosion started a collapse of the gas and dust cloud that eventually formed our sun. The shock waves from the supernova spread throughout the nebula. Gases and other matter continued to spin around the newly formed sun, forming planets, satellites, moons, and asteroids.

Accept all logical speculations about the formation of Pluto. As the newly formed planets began to cool, smaller clumps of matter formed around them. These smaller clumps became moons, or satellites. Many astronomers believe that one of the satellites near Neptune may have broken away from that planet to become Pluto.

If you are teaching thematically, you may want to use the Connections feature to reinforce the themes of systems and interactions or patterns of change.

Integration: Use the Connections feature to integrate astonomy into your lesson.

out when discussing development of a new species that a rapid or sudden change takes place over thousands of years.

Explain that Gould and Eldridge believed that species live in a state of equilibrium or balance for many years. Then something interrupts or punctuates this state of equilibrium causing the species to change. Explain that many scientists today conclude that both theories are partly correct. They believe that evolution is a combination of gradualism and punctuated equilibrium.

INDEPENDENT PRACTICE

Section Review 2–4

1. Punctuated equilibrium is a theory that there may be periods in Earth's history in which many adaptive radiations occurred in a relatively short time. A mass extinction is the relatively simultaneous extinction of a great many species.

Many adaptive radiations occur after a mass extinction. Thus, mass extinctions may be one cause of the sudden adaptive radiations that are part of the theory of punctuated equilibrium.

2. The fossil record shows very little evidence of gradual change over time. It does, however, show sudden adaptive radiations.

REINFORCEMENT/RETEACHING

Review students' responses to the Section Review questions. Reteach any material that is still unclear, based on their responses.

CLOSURE

▸ *Review and Reinforcement Guide*

Have students complete Section 2–4 in the *Review and Reinforcement Guide.*

Laboratory Investigation

ANALYZING A GEOLOGIC TIME LINE

BEFORE THE LAB

1. Gather all materials at least one day prior to the investigation. You should have enough supplies to meet your class needs, assuming six students per group.
2. Adding machine tape is generally available in large rolls in office supply and stationery stores.

PRE-LAB DISCUSSION

Before beginning this activity, review with students the concept of a scale of distance, and relate this concept to that of a time scale, in which a fixed unit of distance on the scale corresponds to a given amount of time, rather than to another distance. Point out that using such a scale to locate events in time is quite similar to using a labeled graphical axis to plot points. Also, review briefly the two most important divisions of the geologic time scale—the era and the period. Refrain from actually comparing the lengths of the different eras or periods, although you may wish to have students speculate on their relative lengths in terms of mathematical factors.

Laboratory Investigation

Analyzing a Geologic Time Line

Problem

How can the relationships between evolutionary events be plotted on a time line?

Materials (per student)

meterstick
pencil
5 meters of adding-machine tape

Procedure

1. For your time line, use a scale in which 1 mm = 1 million years, or 1 m = 1 billion years.
2. Using the meterstick, draw a continuous straight line down the middle of the tape. Draw a straight line across one end. Label this line *The Present*. Assuming each meter represents 1 billion years, place a label at the spot representing 4.6 billion years ago. Add the label *Earth's Beginning?* to this line.
3. Using the table provided, plot each event on the time-line tape. Label both the number of years ago and the event.

Observations

1. Which time period is the longest? The shortest?
2. In which era did dinosaurs exist or begin to exist? In which era did mammals exist or begin to exist?
3. Which lived on Earth the longer time, dinosaurs or mammals?

Analysis and Conclusions

1. How many years does 1 cm represent on your time scale?
2. Why is there a question mark in the label Earth's Beginning?
3. If you had any difficulty plotting some of the events on the list, explain why.
4. What general conclusions can you draw from your time line regarding gradual evolution versus punctuated equilibrium?
5. **On Your Own** Add any events to your time line that you feel are significant. Use the information you have read in both Chapters 1 and 2.

INFERRED AGES OF EVENTS IN YEARS BEFORE PRESENT	
Event Label	**Number of Years Ago**
First mammals and dinosaurs	200 million
Beginning of Carboniferous Period	345 million
Oldest fungi	1.7 billion
Beginning of Jurassic Period	190 million
Beginning of Devonian Period	395 million
Last ice age	10,000
Beginning of Cretaceous Period	136 million
Beginning of Paleozoic Era (first abundant fossils)	570 million
Oldest rocks known	4.2 billion
Beginning of Quaternary Period	1.8 million
Oldest carbon from plants	3.6 billion
Beginning of Ordovician Period	500 million
First birds	160 million
Beginning of Cenozoic Era	65 million
First humanlike creatures	2–4 million
Beginning of Silurian Period	430 million
First reptiles	290 million
Beginning of Mesozoic Era	225 million
Humans begin to make tools	500,000
Beginning of Permian Period	280 million

TEACHING STRATEGY

1. Circulate through the room, assisting those students who are having difficulty with the mathematical conversions.
2. Have the teams follow the directions as they work.

DISCOVERY STRATEGIES

Discuss how the investigation relates to the chapter ideas by asking open questions similar to the following.
• **Which event on the chart is the oldest? Which event happened most recently?**

(Earth's Beginning at 4.6 billion years; last ice age at 10,000.)
• **The written history of human activity is only several thousand years old. Why would it be difficult to add important historical events such as the signing of the Declaration of Independence to your time line?** (Human written history occupies an extremely short length of time relative to the geologic history of the Earth. There would not be enough space on the time line to write many events.)
• **About how much space would your life**

span so far occupy on your time line? (A 13-year-old student's life span would occupy only 13 millionths of a millimeter.)

OBSERVATIONS

1. The Paleozoic Era is the longest time period. The Quaternary Period is the shortest time period.
2. Both dinosaurs and mammals began to exist in the Mesozoic Era.
3. Based on the chart, they appeared around the same time. The earliest mammals did precede the dinosaurs.

Study Guide

Summarizing Key Concepts

2–1 Evolution: Change Over Time

▲ Evolution can be defined as a change in species over time.

▲ A species is a group of organisms that share similar characteristics and can interbreed to produce fertile offspring.

▲ Scientists use fossil evidence, anatomical evidence, embryological evidence, chemical evidence, and molecular evidence to demonstrate that evolution has occurred during Earth's 4.6-billion-year history.

▲ By studying homologous structures, Lamarck demonstrated that living things may share common ancestors. Homologous structures are structures that have evolved from the same body parts.

2–2 Charles Darwin and Natural Selection

▲ Based on his observations of organisms in the Galapagos, Charles Darwin developed a theory of evolution that described evolutionary changes as a result of natural selection.

▲ Natural selection is the survival and reproduction of those organisms best adapted to their surroundings.

2–3 The Development of a New Species

▲ In general, organisms that share the same niche must compete with one another. Organisms that occupy separate niches do not compete.

▲ New species evolve when there are empty niches that can be filled or when a species moves into a niche it did not previously occupy.

▲ Speciation, or the development of a new species, may occur when an organism becomes adapted to a new niche.

▲ The process by which one species evolves into several species, each of which fills a different niche, is called adaptive radiation.

2–4 Punctuated Equilibrium

▲ According to the punctuated equilibrium theory, there may be periods in Earth's history in which many adaptive radiations occur in a relatively short period of time. Many of these great adaptive radiations occurred after a mass extinction in which many niches were left unoccupied.

Reviewing Key Terms

Define each term in a complete sentence.

2–1 Evolution: Change Over Time
evolution
adaptation
homologous structure
molecular clock

2–2 Charles Darwin and Natural Selection
natural selection

2–3 The Development of a New Species
niche
adaptive radiation

2–4 Punctuated Equilibrium
punctuated equilibrium

Part 1

You may wish to have interested students construct a more complete geologic time line. Students can do this by first doing library research to obtain more specific information on approximate first appearances and extinctions of a number of organisms and then plotting these time data into the tape time line.

Part 2

Ask students to consider what would happen to the lengths of the intervals involved if they were to attempt to use the 5-m-long tape time scale to plot historical events in human history, without changing the billion-year calibrations. (The students should come to the conclusion that even remote historical events—the building of the pyramids, for example—would need to be plotted extremely—and impracticably—close to the end of the tape labeled "the present." Thus, all of recorded human history represents only a tiny fraction of the total time that has elapsed since the Earth was formed.

ANALYSIS AND CONCLUSIONS

1. 10 million years.

2. The date of the Earth's beginning is open to doubt and is known only approximately.

3. Students will have considerable difficulty in plotting events of the past 2 million years, which cluster together near the end of the tape. The Cenozoic Era (and, in particular, human history) is very short in comparison with the Earth's history.

4. Accept all logical, well-written responses. Many students may suggest that the time line supports punctuated equilibrium because it implies gaps in the fossil record.

5. Student responses will vary.

Chapter Review

Chapter Review

ALTERNATIVE ASSESSMENT

The *Prentice Hall Science* program includes a variety of testing components and methodologies. Aside from the Chapter Review questions, you may opt to use the Chapter Test or the Computer Test Bank Test in your *Test Book* for assessment of important facts and concepts. In addition, Performance-Based Tests are included in your *Test Book*. These Performance-Based Tests are designed to test science process skills, rather than factual content recall. Since they are not content dependent, Performance-Based Tests can be distributed after students complete a chapter or after they complete the entire textbook.

CONTENT REVIEW

Multiple Choice

1. b
2. c
3. b
4. b
5. a
6. b

True or False

1. F, Evolution
2. F, more
3. T
4. F, species
5. T
6. F, isolation
7. T

Concept Mapping

Row 1: Mutation, Species
Row 2: Homologous structures, similar embryo development, similarities in DNA, similar protein structures

Content Review

Multiple Choice

Choose the letter of the answer that best completes each statement.

1. A term that can be described as descent with modification is
   a. natural selection.
   b. evolution.
   c. isolation.
   d. migration.
2. Which of these is not used as evidence of evolution?
   a. fossils
   b. embryology
   c. niche
   d. homologous structures
3. A bat's wing and a lion's leg bones are examples of
   a. variation.
   b. homologous structures.
   c. migratory effects.
   d. fossils.

4. A change that increases an organism's chances for survival is called a(n)
   a. mutation.
   b. adaptation.
   c. radiation.
   d. homologous structure.
5. Another way to say survival of the fittest is
   a. natural selection.
   b. overproduction.
   c. adaptive radiation.
   d. mutation.
6. The process in which one species evolves into several species is called
   a. mutation.
   b. adaptive radiation.
   c. isolation.
   d. punctuated equilibrium.

True or False

If the statement is true, write "true." If it is false, change the underlined word or words to make the statement true.

1. Natural selection can be defined as descent with modification.
2. Most species produce fewer young than the environment can support.
3. Most members of a species show variation.
4. Evolution can be defined as a change in an organism over time.
5. Migration and isolation are two common ways an organism may move into a new or empty niche.
6. The unusual organisms living in Australia are due to migration.
7. Homologous structures are evidence of adaptive radiation.

Concept Mapping

Complete the following concept map for Section 2–1. Refer to pages F6–F7 to construct a concept map for the entire chapter.

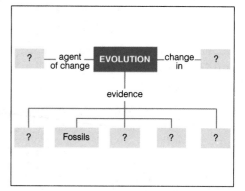

CONCEPT MASTERY

1. Two organisms have evolved from the same ancestor.
2. Through overproduction, nature assures that at least some of the offspring will survive to continue the species.
3. Student examples will vary. In the case of tadpoles, there is competition within the pond for food. Only those tadpoles strong enough to obtain food and fast enough to avoid predators will live and reproduce.

4. Migration is the movement of animals away from their original homes to new places. Isolation occurs when numbers of a species or a group of species are separated from the rest of their kind for long periods of time.
5. In general, evolution is a slow process in which changes are not particularly obvious over short periods of time. You would not likely see evolutionary changes during a single lifetime. Even punctuated equilibrium usually takes longer than the human life span.

Concept Mastery

Discuss each of the following in a brief paragraph.

1. What does the phrase common descent mean?
2. Discuss the role of overproduction in nature.
3. Use an example to explain the concept of natural selection.
4. Compare migration and isolation.
5. Evolution is an ongoing process. It continues as it has for millions of years. Why are scientists usually unable to see evolution in action?
6. Compare natural selection as described by Darwin and punctuated equilibrium as described by Gould and Eldridge.

Critical Thinking and Problem Solving

Use the skills you have developed in this chapter to answer each of the following.

1. **Making comparisons** Would you expect there to be more similarities between the DNA of a cat and a lion or a cat and a dog? Explain.
2. **Relating cause and effect** Certain snails that live in woods and in grasses are eaten by birds. The snails that live in grasses are yellow. The snails that live on the woodland floor are dark colored. Explain how the snails have become adapted to their environments through natural selection.
3. **Making observations** Observe an animal in your classroom, your home, or a pet store. List five characteristics of the animal, such as hair color or size. Then list possible variations for each characteristic. Finally, explain how each variation might make an animal more fit for survival in its natural environment.
4. **Making inferences** Darwin was amazed at the diversity of life he observed on the Galapagos. How might this diversity have contributed to his ideas regarding evolution?
5. **Evaluating options** Is protecting an endangered species defying natural selection? Explain your answer.

6. **Applying concepts** The giant panda occupies a very small niche by eating only one kind of food: bamboo. How can being adapted to such a small niche actually endanger this species?

7. **Using the writing process** You are a young reporter for a local newspaper near the home of Charles Darwin. You have been asked to interview Darwin about his theory of evolution. Develop a list of five questions you would like to have Darwin discuss. Then see whether you can answer them in the manner Darwin would.

---

5. Accept all logical responses. Students may point out that natural selection requires a certain minimum number of individuals in a species in order to operate. Also, many endangered species have become endangered through human activity rather than through "natural" events.
6. A change in the environment that threatened the bamboo plants would also threaten the giant panda.
7. Student questions should be consistent with the information presented in the text.

KEEPING A PORTFOLIO

You might want to assign some of the Concept Mastery and Critical Thinking and Problem Solving questions as homework and have students include their responses to unassigned questions in their portfolio. Students should be encouraged to include both the question and the answer in their portfolio.

ISSUES IN SCIENCE

The following issues can be used as springboards for discussion or given as writing assignments.
1. Have students debate the following statement: "The extinction of a species due to human destruction of its natural environment is an unfortunate but unavoidable result of modern progress."
2. Based on students' knowledge of natural selection, have them explain why some insecticides, such as DDT, are no longer effective against certain species of insects. Does this imply that we should invent new and better insecticides? Or are there other solutions to the threats that insects pose to agriculture and humans?

---

6. Natural selection as described by Darwin is a long, slow process in which change occurs at a relatively constant rate. In punctuated equilibrium, periods of no change are followed by periods of rapid change.

CRITICAL THINKING AND PROBLEM SOLVING

1. There would be more similarities between a cat and a lion because they are more closely related and are in the same family.

2. The woodland floor is usually dark, and dark-colored snails would be less easily seen by predators. In the same way, yellow snails would be less easily seen in grasses than dark-colored snails.
3. Observations and conclusions should be consistent with the selected animal.
4. Each of the many organisms observed by Darwin seemed perfectly adapted to survival in its particular environment. Darwin's theories helped to explain why such a great diversity of life exists.

SECTION	HANDS-ON ACTIVITIES

3–1 The Search for Human Ancestors
pages F78–F81
 	
Multicultural Opportunity 3–1, p. F78
ESL Strategy 3–1, p. F78
	Student Edition
	  ACTIVITY (Discovering): A Modern Culture, p. F80
	  LABORATORY INVESTIGATION: Comparing Primates—From Gorillas to
	     Humans, p. F92
	
	Activity Book
	  CHAPTER DISCOVERY: Analyzing Primate Proteins, p. F77
	
	Teacher Edition
	  Comparing Primates, p. F76d
	  Flexible Hands, p. F76d
	  Opposing Thumbs, p. F76d

3–2 Human Ancestors and Relatives
pages F82–F91
 	
Multicultural Opportunity 3–2, p. F82
ESL Strategy 3–2, p. F82
	Student Edition
	  ACTIVITY (Discovering): Communication, p. F85
	
	Laboratory Manual
	  A Human Adaptation, p. F39

Chapter Review pages F92–F95	

OUTSIDE TEACHER RESOURCES

Books

Brace, C. *The Stages of Human Evolution: Human and Cultural Origins*, 2nd ed., Simon & Schuster.

Emergence of Man, 20 vols., Time-Life Books.

Gould, S. *Ever Since Darwin*, W. W. Norton.

Leakey, R. *Human Origins*, Lodestar/ Dutton.

Reader, J. *Missing Links: The Hunt for Earliest Man*, Little, Brown.

Sattler, Helen Roney. *Hominids: A Look Back at Our Ancestors*, Lothrop.

Szalay, F., and E. Delson. *Evolutionary History of the Primates*, Academic Press.

Zihlman, A. *The Human Evolution Coloring Book*, Harper Row.

Audiovisuals

Evolution of Man, film, Coronet

Inquiry Into Prehistoric Life, sound filmstrips, Encyclopaedia Britannica Education

Leakey, video or film, National Geographic

Monkeys, Apes, and Man, video or film, National Geographic

OTHER ACTIVITIES	MEDIA AND TECHNOLOGY
Activity Book ACTIVITY: A Puzzling Find, p. F85 ACTIVITY: Koko and Her Kitten, p. F87 **Review and Reinforcement Guide** Section 3–1, p. F25	**English/Spanish Audiotapes** Section 3–1
Student Edition ACTIVITY (Writing): The Iceman, p. F90 **Activity Book** ACTIVITY: An Evolutionary Chart, p. F81 ACTIVITY: Shipwrecked! p. F83 **Review and Reinforcement Guide** Section 3–2, p. F27	**English/Spanish Audiotapes** Section 3–2
Test Book Chapter Test, p. F53 Performance-Based Tests, p. F71	**Test Book** Computer Test Bank Test, p. F59

*All materials in the Chapter Planning Guide Grid are available as part of the Prentice Hall Science Learning System.

CHAPTER OVERVIEW

The topic of human origins and evolution is probably one of the most controversial and changeable topics in science. Although people in Darwin's time were offended by the notion that humans evolved, most people today accept the idea that humans are members of the primate order and arose from apelike ancestors. Now the controversy centers on how humans have evolved. With the discovery of each new type of hominid fossil, theories of human evolution are revised. At any one time, there are usually several rival evolutionary schemes to consider.

The primates share several characteristics including flexible fingers and three-dimensional vision. Eventually the primates gave rise to an evolutionary line called hominoids, which include the apes and humans. The hominoids, in turn, gave rise to a group of species called the hominids. Scientists recognize the hominids as the closest relatives of modern-day humans.

Hominids displayed many characteristics that distinguished them from other hominoids. Among these characteristics are the physical attributes that enable all hominids to walk erect. The earliest human ancestor is *Homo habilis,* which means skillful human. The closest ancestor, the Cro-Magnon, is classified as *Homo sapiens sapiens* just as present-day humans are. They are the first hominids truly identical to humans. It is believed that the Cro-Magnon communicated with one another through language. Complex language is one characteristic that distinguishes humans from other primates.

3–1 THE SEARCH FOR HUMAN ANCESTORS
THEMATIC FOCUS

The purpose of this section is to focus on the characteristics of the primates in general and the evolutionary phases that gave rise to hominids, the closest relatives of humans. Students will learn that scientists agree that humans share common ancestors with other primates such as chimpanzees. Students learn that as a group, primates display certain important adaptations such as three-dimensional vision, flexible fingers, and large and complex brains. Most also have opposable thumbs that enable them to grasp objects more easily.

Students will discover that early in their history, primates split into several evolutionary lines. One of these lines was the anthropoids, which gave rise to the two major groups of monkeys—Old World monkeys and New World monkeys. The Old World monkeys eventually gave rise to the group called hominoids, from which humans are descended.

The themes that can be focused on in this lesson include evolution, patterns of change, scale and structure, and unity and diversity.

*Evolution: Primates are relative newcomers on Earth. Animals have inhabited Earth for more than 600 million years, but the first known primates appeared about 70 million years ago. About 50 million years ago, the primates evolved into two separate paths—prosimians and anthropoids. Anthropoids also evolved into separate paths, giving rise to the hominids, the closest relatives of present-day humans.

*Patterns of change: As primates evolved, they developed certain characteristics that enabled them to survive in their environment. These characteristics include flexible hands and three-dimensional vision. The hands made it easier for primates to grasp objects, and the three-dimensional vision gave primates depth perception.

Scale and structure: The body structures of primates vary according to their life functions. Humans, for example, have neck and pelvic bones that make walking erect possible. The ability to walk erect, in turn, freed the use of hands for tasks other than walking.

*Unity and diversity: Although each group of primates has different structures, they all have certain basic characteristics in common. These characteristics include flexible hands, flattened faces, eyes pointing forward, and three-dimensional vision.

PERFORMANCE OBJECTIVES 3–1

1. **Explain how scientists use fossils to trace the path of human evolution.**
2. **Describe the characteristics common to all primates.**
3. **Compare early primates and modern humans.**

SCIENCE TERMS 3–1

primate, p. F78

3–2 HUMAN ANCESTORS AND RELATIVES
THEMATIC FOCUS

The purpose of this section is to describe the characteristics of hominids and to explain how trends in their evolution gave rise to modern humans. Students learn that hominids displayed several characteristics that distinguished them from other hominoids. Hominids were able to walk erect and had much larger brains than other hominoids.

Students learn that most evidence for hominid evolution comes from fossils found in Africa. They identify the first hominids as members of the genus *Australopithecus.* Later hominids, more closely related to humans, belong to the genus *Homo.* These species include *Homo habilis, Homo erectus,* and early *Homo sapiens.* Modern humans are classified as *Homo sapiens sapiens.*

Students find out that the first hominids very similar to modern humans were the Cro-Magnons, who are classified as *Homo sapiens sapiens* just as humans are. The Cro-Magnons developed an oral language for communicating with one another.

The themes that can be focused on in this section include energy, systems and interactions, and stability.

Energy: Hominids, like other primates, obtain their energy by eating plants and other animals. Neanderthals were experts at controlling fire and probably used the heat energy from the fire to cook foods and keep warm.

*Systems and interactions: Primates respond to and interact with their environment in ways that help them gather food, reproduce, and protect themselves. Evidence indicates that the Neanderthal had a belief system about their world and nature. One way Cro-Magnons interacted with one another was through language.

Stability: The various life functions of primates help to maintain a stable internal and external environment.

PERFORMANCE OBJECTIVES 3–2

1. **Discuss the use of fossil record in attempting to find the earliest human.**
2. **Compare the use of fossil evidence to the use of chemical evidence in determining the path of human evolution.**
3. **Identify characteristics of different ancestors of humans.**
4. **Compare *Homo habilis* and *Homo erectus*.**
5. **Describe some characteristics associated with the way of life of Neanderthals and Cro-Magnons.**

SCIENCE TERMS 3–2

Neanderthal, p. F88
Cro-Magnon, p. F89

Discovery *Learning*

TEACHER DEMONSTRATIONS MODELING

Comparing Primates

Have students compare a photograph of a chimpanzee obtained from a nature or wildlife magazine with one of a human being.
• **What characteristics do these two organisms have in common?** (Accept all logical answers.)
• **How are these organisms different from each other?** (Accept all logical answers.)
• **What adaptations does a chimpanzee have that allow it to live successfully in its natural environment?** (Accept all logical answers.)
Tell students that both the chimpanzee and the human belong to a large group of mammals called the primates. Point out to students that the DNA of human beings is almost 98 percent identical to the DNA of chimpanzees. Evidence suggests that both monkeys and humans have evolved from a common primate ancestor.

Flexible Hands

Ask students to place their hands in front of them and spread their fingers apart. Point out that this is called divergent movement. Students can also try performing divergent movements with their toes.

Next ask students to cup their hands, as if they were filled with water. Point out that this is called a convergent movement. Then ask students to wrap their fingers around objects such as rulers or chalk board erasers. Point out that this is called prehensile movement. Finally, have students hold a pen or pencil in their hands as if they were about to write. Point out that this type of movement, in which the thumb is touching one or more of the tips of the fingers, is called opposable movement.

Explain to students that divergent movements, convergent movements, prehensile movements, and opposable movements are all possible because clawed paws gave rise to hands among certain primates. Also explain that as humanlike creatures evolved into modern humans, the hand bones became modified and these four types of movements became more and more refined.

Opposing Thumbs

Show students the palms of your hands with your fingers and thumb extended. Fold your thumb into the palm and back out several times. Touch each of your fingers in turn with the thumb of that hand. Form your hand into a fist and open it several times. Pick up an object.
• **Can all of you do the exercises I just did?** (Yes.)
• **Is the ability to do these exercises a characteristic shared by most animal species?** (No.)
• **What skills are possible only with hands that have opposing thumbs?** (Accept all logical answers.)
• **What skills are possible with hands that do not have opposing thumbs?** (Accept all logical answers.)
• **How would an opposing thumb be of use to a monkey or chimpanzee?** (Accept all logical answers.)

CHAPTER 3
The Path to Modern Humans

INTEGRATING SCIENCE

This life science chapter provides you with numerous opportunities to integrate other areas of science, as well as other disciplines, into your curriculum. Blue numbered annotations on the student page and integration notes on the teacher wraparound pages alert you to areas of possible integration.

In this chapter, you can integrate life science and primates (p. 78), earth science and plate tectonics (p. 80), geography (p. 83), life science and blood (p. 84), language arts (pp. 88, 90), social studies (p. 89), and physical science and electromagnetic radiation (p. 91).

SCIENCE, TECHNOLOGY, AND SOCIETY/COOPERATIVE LEARNING

The Earth and the organisms that inhabit it have evolved over millions of years to their present state. But what if you could control evolution? What would you change? Would humans still be the most intelligent life form? Today, there is a computer simulation "game" that allows the player to control evolution by simulating the life cycle of a planet. The player can cause mass extinctions, change biomes from one type to another, create an atmosphere of any composition, and control the evolution of all the organisms on the planet.

This advanced computer simulation puts the user in control of evolution on

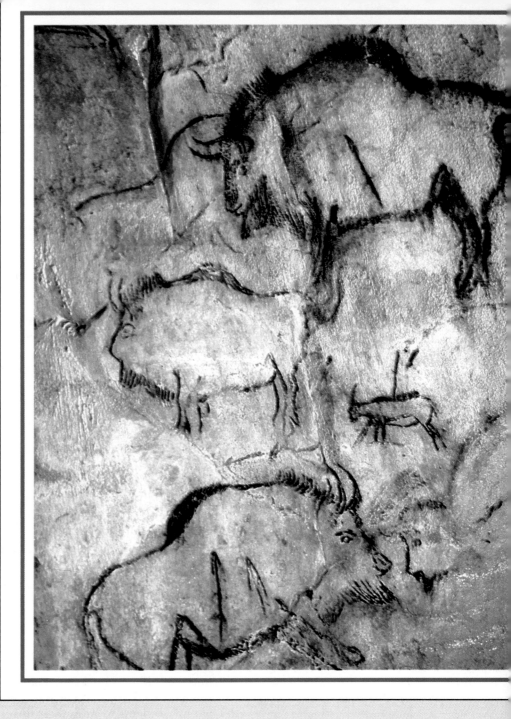

INTRODUCING CHAPTER 3

DISCOVERY LEARNING

▶ *Activity Book*

Begin your teaching of the chapter by using the Chapter 3 Discovery Activity from your *Activity Book*. Using this activity, students will discover which primates are most closely related to humans based on analysis of the amino acid sequences in hemoglobin.

USING THE TEXTBOOK

Have students observe the chapter-opening photograph and caption.
- **When do you think these drawings were made?** (No information is given about when the drawings were made; students may surmise, however, that the primitive nature of the drawings indicates that they were drawn many thousands of years ago.)

You might display pictures of present-day animals that have characteristics similar to those portrayed in the drawings.

- **Which modern-day animals most closely resemble those shown in the drawing?** (Students might name the American bison, the bull, and the deer.)

Tell students that the drawings were made about 15,000 years ago.
- **Why might the artist have chosen to draw these animals?** (Students may suggest that the animals in the ecosystem or the value of the animals in meeting the needs of the people made the animals ideal subjects for an artist's portrayal.)

The Path to Modern Humans

Guide for Reading

After you read the following sections, you will be able to

3–1 The Search for Human Ancestors
- Describe the characteristics of primates.
- Compare New World monkeys to Old World monkeys.

3–2 Human Ancestors and Relatives
- Describe some early human ancestors.
- Compare Neanderthals and Cro-Magnons.

In 1879, on a farm in northern Spain, a twelve-year-old girl named Maria accidentally made the first discovery of prehistoric art. While exploring a cave with her father, she discovered some remarkable drawings. These drawings pictured deer, wolves, and large bull-like animals. In addition, Maria and her father found stone tools and animal bones—objects that were depicted in the paintings.

Unfortunately, the archaeologists who were told of the discovery dismissed it as a fraud. You see, most archaeologists of the 1800s believed that ancient people had neither the ability nor the intelligence to create such works of art.

Archaeologists today believe that the animals in the cave were painted by skilled artists who lived about 15,000 years ago. Caves with similar paintings have also been found in other parts of Europe. In this chapter you will learn how scientists have used these discoveries to piece together the story of human history. And perhaps one day you will be as lucky as Maria and discover something equally exciting!

Journal *Activity*

You and Your World Focus on your state and predict what the environment will be like in 10,000 years. In your journal, describe the climate, geographic features, composition of the air, type of vegetation, and available natural resources. Make a drawing of your new environment and include a few animals of your own design.

◀ *These paintings of bison in a cave in France are similar to those discovered in Spain in 1879.*

F ■ 77

planet Earth. This computer program uses advanced technology, so that its models for atmosphere, climate, and biosphere work together to generate continuously updated data. The player guides the planet's evolution into decline or into a higher civilization by creating hostile environments and cataclysmic events to see which species of organisms are able to evolve and survive. For example, if the player increases the volcanic activity level, there will be a corresponding change in the atmosphere. This event, in turn, causes evolution among the organisms on the planet. What would happen if you were in control of evolution on Earth? Would people have evolved in the same manner, or would dinosaurs have become the most intelligent organisms on Earth?

Cooperative learning Using preassigned lab groups or randomly selected teams, have groups complete the following assignment.

Have each group generate a description of a virgin planet. Their descriptions should include the temperature, type of atmosphere, number and location of landmasses and bodies of water, position in the solar system, and any other pertinent data. When groups have finished their descriptions, read the descriptions aloud and have students select one description that they will use as the virgin planet on which they are in control of evolution. Final products could be a mural that shows the changes that have occurred, a news documentary, or any other form that you approve.

See Cooperative Learning in the *Teacher's Desk Reference.*

JOURNAL ACTIVITY

You may want to use the Journal Activity as a basis of discussion. Have students consider whether they think humans will listen to or ignore environmental warnings before they begin writing. Students' journal entries should then reflect their positions on the issue of protecting the environment. Students who believe that environmental warnings may not be heeded might want to represent mutant forms of life adapted to new environmental conditions. Students should be instructed to keep their Journal Activity in their portfolio.

Have students read the chapter-opening text.
- **Why did archaeologists dismiss the findings of Maria and her father?** (They believed that early humans were incapable of artistic endeavors.)
- **Why do you think attitudes have changed about the drawings?** (Answers will vary but should reflect the understanding that additional evidence corroborates the findings of Maria and her father.)

- **What do the drawings tell us about early humans?** (They were skilled artisans aware of their surroundings and animals in their ecosystems.)
- **Why might people be interested in learning about early humans and their origins?** (Accept all logical answers; you might relate such study to tracing family histories to better understand who we are.)

3–1 The Search for Human Ancestors

Ecosystem destruction in South America and Africa is having a serious effect on the jungle primates in those areas. Have your students investigate some of the efforts to protect these organisms. You might have them investigate especially the life and work of Jane Goodall and Dian Fossey, two prominent researchers dedicated to the preservation of primate environments.

ESL STRATEGY 3–1

Explain the meaning of chronological order. Ask for the meaning of alphabetical order (an area in which LEP students often need practice). Have students place the following words first in chronological order and then in alphabetical order. Also, have them list the major characteristics of primates.

primates—mammals—animals

Have students circle the word in each group that is incorrectly used and insert the corrected form.
- lemurs, lorises, anos-anos
- monkeys, snakes, humans

Then ask them to make a chart listing each primate group. The headings should be: Kingdom, Phylum, Class, Order, Family, Genus, and Characteristics.

Guide for Reading

Focus on these questions as you read.
▶ What are some characteristics of primates?
▶ How did primates evolve?

Figure 3–1 *Archaeologists are scientists who study the remains of ancient people. These archaeologists are examining some specimens at a site in Chichen-Itza, Mexico.*

78 ■ F

3–1 The Search for Human Ancestors

The search for the first ancestors of humans is on. All over the Earth, scientists are piecing together details that will shed more light on the exciting study of our past. Anthropologists and archaeologists examine ancient human tools and cultures for answers. Paleontologists study the fossils of ancient humans and compare them with living forms. Biologists examine the DNA (basic substance of heredity) of different species, looking for similarities and differences that determine whether the species are closely related.

Each new fossil find and each new research paper brings a new storm of controversy. Just when scientists think that years of study and debate have brought them a step closer to human origins, a new fossil or a new theory turns up. Then a fresh shadow of doubt is cast on our knowledge of how humans evolved.

There is no doubt among scientists, however, that humans evolved from common ancestors they share with other living primates. Scientists also know that the human species evolved in Africa and then spread around the Earth.

What Are Primates?

Primates are members of a group of mammals that include humans, monkeys, and about 200 other species of living things. **Primates share several important characteristics. All primates have flexible fingers. (Some have flexible toes, too.) Most primates have opposable thumbs.** An opposable thumb is opposite the other fingers and is able to move toward them and touch them. Opposable thumbs enable primates to grasp objects, both large and small.

Generally, primates have much flatter faces than other groups of mammals. Their eyes are located at the front of their heads rather than at the sides, and their snouts are very much reduced in size. As a result of these features, the brains of primates can ①

TEACHING STRATEGY 3–1

FOCUS/MOTIVATION

Display a picture of a zebra and a gorilla.
- **How are these animals different from each other?** (Accept all logical answers.)
- **How is the placement of the gorilla's eyes different from that of the zebra?** (In the gorilla, the eyes are closer together

and are found at the front of the head rather than at the sides.)
- **Which animal would you expect to have the wider field of vision?** (The zebra, because its eyes are set on the sides of its head.)

Explain that each eye perceives a different image.
- **In which animal would the images from the eyes overlap the most? How can you tell?** (The gorilla, because both eyes face

in the same direction.)

Tell students that the placement of the gorilla's eyes allows its brain to combine the images from each eye into one three-dimensional picture. The ability of the eyes to form a three-dimensional picture is called stereoscopic vision.

Instruct students to look straight ahead and close or cover one eye.
- **What happens?** (Accept all answers. Students should discover that objects

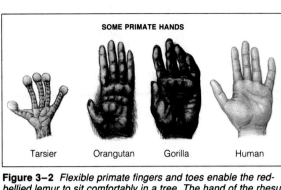

SOME PRIMATE HANDS

Tarsier Orangutan Gorilla Human

Figure 3–2 *Flexible primate fingers and toes enable the red-bellied lemur to sit comfortably in a tree. The hand of the rhesus monkey, which is shown on top of a human hand, has shortened fingers. How are the hands of the tarsier, orangutan, and gorilla similar to the human hand? How are they different?* ❶

combine the separate image from each eye into a single, three-dimensional picture. **The ability to form a three-dimensional picture is another characteristic of primates.** Thus, primates can sense depth and can judge distances. This adaptation is especially helpful when a primate has to locate branches as it swings from one tree to the next.

Primates also have a large and complex cerebrum. The cerebrum is the part of the brain responsible for all the voluntary actions of the body. The behavior of primates is therefore more involved than that of any other animal. For example, primate mothers take care of their young for a longer period of time than most other mammals do. Many species of primates also have complicated social behaviors that include friendships and—sadly—fighting among competing groups.

A New Kind of Primate

On the evolutionary time scale, primates are considered newcomers. Animals have inhabited Earth for more than 600 million years. Mammals have been here for at least 200 million years. In contrast, the first known primates appeared about 70 million years ago—relatively recently.

Some 50 million years ago, the early primates split into two main evolutionary groups—prosimians (proh-SIHM-ee-ehnz) and anthropoids (AN-thruh-poidz). Modern prosimians are almost all nocturnal animals with large eyes adapted for seeing in the

Figure 3–3 *What primate characteristic is shown in this photograph of a tamarin?* ❷

appear flat, or two dimensional, when they look through only one eye.)

• **Do humans have stereoscopic vision?** (Yes.)

• **Why is stereoscopic vision important?** (Accept all logical answers. Guide students to understand that stereoscopic vision enables humans to perceive depth and judge distances.)

A MODERN CULTURE

Discovery Learning

Skills: Relating concepts, applying concepts, making judgments, making inferences

Materials: objects of students' choice

This activity helps students learn to think objectively about their culture and its attitudes, values, and artifacts.

Answers

1️⃣ Hominoids include apes—gibbons, orangutans, gorillas, and chimpanzees—and humans.

Integration

1️⃣ Earth Science: Plate Tectonics. See *Dynamic Earth*, Chapter 3.

FACTS AND FIGURES

THE ORANGUTAN

The orangutan is the only great ape to live outside of Africa. It lives exclusively in the forests of Borneo and Sumatra. When it stands erect, the orangutan is about 1.4 meters tall, and it has an arm span that stretches more than 2 meters.

3–1 (continued)

Primates are divided into two main types: anthropoids—humans, apes, and monkeys; and prosimians—lemurs, tarsiers, lorises, and pottos. Although primates have a number of characteristic physical features, not every primate has each of these features. The anthropoids have more of the characteristic primate features than do the prosimians. For the most part, the anthropoids are larger in size and have relatively larger and more complex brains than do the prosimians.

ACTIVITY
DISCOVERING

A Modern Culture

Prepare a display that contains some objects that represent the culture in which you live. Choose objects that would help a scientist 10,000 years from now understand your culture.

■ Explain why you chose the objects you did.

Figure 3–4 *This evolutionary tree illustrates how primates may have evolved from early prosimians. Which primates are hominoids?* 1️⃣

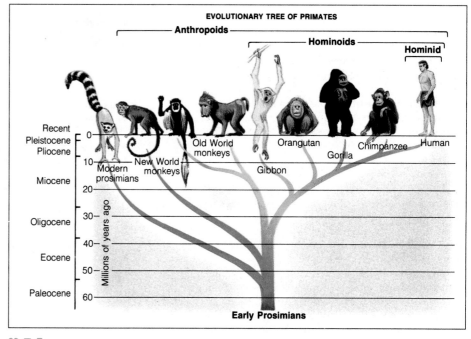

EVOLUTIONARY TREE OF PRIMATES

80 ■ F

---

dark. Some examples of living prosimians are lemurs, lorises, and ayes-ayes. Anthropoids are humanlike primates that include monkeys, apes, and humans.

A few million years after the split of prosimians and anthropoids, two anthropoid branches evolved. These branches, the two major groups of monkeys and apes, developed when the continents separated 1️⃣ from one another. One group of anthropoids evolved into the monkeys found today in Central and South America. This group is called the New World monkeys. (The term New World comes from the days of Columbus, when the Americas were referred to as the New World.) All New World monkeys live in trees. Many of them have grasping tails, which help them to move through their habitat. New World monkeys include marmosets, howler monkeys, and spider monkeys.

The other group of anthropoids evolved into Old World monkeys and hominoids (HAHM-eh-noidz). Old

---

Almost all primates except humans live mainly in tropical and subtropical climates. Most primates live in trees, but some spend much of their time on the ground. The primate infant, especially the baby human and ape, depends heavily on its mother for care.

● ● ● ● **Integration** ● ● ● ●

Use the discussion of the separation of the continents to integrate concepts of plate tectonics into your lesson.

Skills Development

Skills: Manipulative, making comparisons, drawing conclusions

To help demonstrate the utility of the opposable thumb, have students pick up a pen and write their names on a sheet of paper. Then have students repeat the activity, but this time ask them not to use their thumbs. Once students have written the names twice, have them compare their signatures.

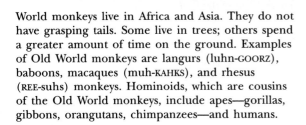

Figure 3–5 *The loris is an example of a modern prosimian (top left). Modern prosimians, almost all of which are nocturnal, are tree-dwelling primates. New World monkeys are also tree-dwelling animals. Some, such as squirrel monkeys (bottom left), can hang from their tails. The macaque is an Old World monkey (right). Old World monkeys do not have grasping tails. They walk on the ground, using all four limbs.*

World monkeys live in Africa and Asia. They do not have grasping tails. Some live in trees; others spend a greater amount of time on the ground. Examples of Old World monkeys are langurs (luhn-GOORZ), baboons, macaques (muh-KAHKS), and rhesus (REE-suhs) monkeys. Hominoids, which are cousins of the Old World monkeys, include apes—gorillas, gibbons, orangutans, chimpanzees—and humans.

3–1 Section Review

1. What are some characteristics of primates?
2. What is three-dimensional vision?
3. How did primates evolve?
4. How are New World monkeys and Old World monkeys the same? How are they different?

Connection—*You and Your World*
5. Design an experiment to demonstrate the advantage of having opposable thumbs.

• **Was it easier to write your names with or without the use of your thumbs?** (Students will find that it was much easier to grasp and control the pen or pencil when using their thumbs.)
• **Which signature is more legible—the first or the second one?** (The first signature should be more legible because the thumb helps control the use of the pen.)
• **Why is the opposable thumb important to primates?** (It enables primates to grasp and control objects.)

ECOLOGY NOTE
THREATENED AND ENDANGERED SPECIES

Three of the hominids are on the threatened or endangered species list. The gorilla of central and western Africa and the orangutan of Borneo and Sumatra are on the endangered species list; the chimpanzee of western and central Africa are on the threatened species list. Suggest that students research information about the habitat in which one of these species lives and find out why its species is in danger.

3–2 Human Ancestors and Relatives

3–2 Human Ancestors and Relatives

About 6 million years ago, the hominoids gave rise to a small group of species now considered to be the closest relatives to humans. The small group of species is called hominids (HAHM-uh-nihdz). Hominids, which include humans and closely related primates, are members of the human family known as Hominidae (hahm-uh-NIHD-igh). Although these early hominids were not yet humans, they did take evolutionary paths that distinguished them from the other hominoids.

The early hominids experienced changes in the shapes of their spinal column and their hip and leg bones. These changes enabled hominids to walk upright on two legs. In this position the hands of hominids were free to use tools more often. At the same time, the opposable thumb evolved, allowing hominids to grasp objects and use them as tools more effectively than other primates did. In addition, hominids showed an unusual increase in brain size. Even for primates, the brains of hominids were exceptionally large.

Figure 3–6 *Compare the skeleton of a human with the skeleton of an ape. Note the shape of the jaws, the structure of the pelvis, and the way the spine enters the skull.*

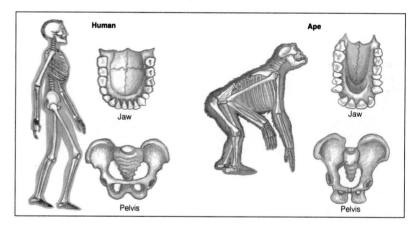

TEACHING STRATEGY 3–2

FOCUS/MOTIVATION

Refer students to Figure 3–6 and ask them to compare the skeleton and skeletal parts of a human to those of an ape.

• **How are the skeletons of the two species alike?** (Basic structure is similar, although the arms of the ape are longer and the legs of the ape are shorter in proportion to body size than those of the

human.)

• **How are the jaws alike and different?** (The jaws have the same number of teeth, but the human jaw is wider and shorter than that of the ape.)

• **How are the pelvises alike and different?** (The general structure is similar, but the human pelvis is wider and shorter than that of the ape.)

CONTENT DEVELOPMENT

Explain that scientists are still unsure about which primate is the earliest ancestor of humans. Some scientists think that *A. africanus* is the earliest ancestor of humans. The first fossil of this species was found in 1924 in a limestone quarry in South Africa. It is believed that this species lived between 2 million and 3 million years ago. Members of this species are thought to have been 1 meter tall. They had small teeth, walked on two legs, and

The First Humanlike Ancestors: *Australopithecus*

In order to determine the course of human evolution, fossils of human ancestors were needed. Much of the evidence for the evolution of hominids came from a small area in eastern Africa between Ethiopia and Tanzania. There, many fossil discoveries of hominids were made.

In 1924, the first fossil of a hominid was discovered in South Africa by the Australian physician Raymond Dart. He named his find *Australopithecus* (AW-struh-loh-pihth-eh-cuhs), which means southern ape. Because the fossil was that of a child hominid, it was of little help to researchers who needed to know what an adult looked like. Fortunately, a few years later in Africa, researchers did find a fossil of an adult hominid. They named it *Australopithecus africanus* (af-rih-KAHN-uhs). *A. africanus* was hailed as the first human ancestor. (The letter *A.* is an abbreviation for the genus *Australopithecus*.) Most fossil specimens of *A. africanus* are only 2 to 3 million years old. The fossils show a blending of apelike and humanlike characteristics that include small teeth, signs of upright posture, and a brain whose size lies somewhere between that of apes and humans.

In 1974, another more complete hominid fossil was discovered in Ethiopia by a team of researchers led by the American scientists Donald Johanson and Tim White. This humanlike fossil was named *Australopithecus afarensis* (af-uh-REHN-zihs), and nicknamed Lucy. The nickname comes from the Beatles' song "Lucy in the Sky With Diamonds," which was playing on the tape deck at the researchers' camp site. The age of this fossil is estimated at about 3.5 million years. Some scientists immediately proclaimed Lucy to be the earliest human ancestor. Other scientists, however, disagreed.

Whether Lucy is one of the earliest human ancestors is still open to question. From other bones found with Lucy, some scientists have concluded that she walked upright on two legs. Others say that her

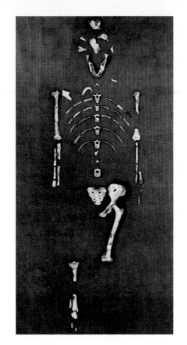

Figure 3–8 *These specimens of* Australopithecus africanus, *found in Taung, South Africa, are about 2.5 million years old.*

BACKGROUND INFORMATION
THE HUMAN ADVANTAGE

Although humans are considered to be primates, they have several characteristics that make them different from the other species in the order. Humans walk exclusively on two legs rather than on four. This advantage is made possible by arches in the feet that help support the entire mass of the body and by a wider pelvis that distributes the body weight more evenly over the two legs. By walking on two feet, humans gained a large advantage—the freedom to use their hands for many tasks. Humans have smaller and less pointed teeth than other primates, resulting in smaller, more recessed mouths and jaws.

The most striking difference between humans and other primate species is the ability of humans to develop complex systems of written and verbal communication. Although other primates have communication techniques, humans have constructed sophisticated languages in both oral and written forms.

had a brain that was larger than an ape's but smaller than a human's.

Other scientists think the oldest humanlike primate was found in Africa in 1977. This primate was called *A. afarensis* and is estimated to be approximately 3.5 million years old. The fossil remains, nicknamed Lucy, were thought to have been that of a female who was roughly 20 years old. She had apelike teeth and jaws and a small brain.

Explain that although Lucy had many apelike characteristics, some scientists believe she was able to walk on two legs. Scientists developed this theory based on the shape of Lucy's pelvis and the way in which the skull sits on top of the spine instead of in front of it.

● ● ● ● **Integration** ● ● ● ●

Use the information about the location of hominid fossils to integrate concepts of geography into your lesson.

In 1925, John T. Scopes went to trial for violating a state law that prohibited teaching evolutionary theories. For two weeks in the heat of the summer, the nation watched as Clarence Darrow, a famed trial lawyer, argued in defense of Scopes in Tennessee. A key witness for the prosecution was William Jennings Bryan, known as the "Great Commoner" for his progressive politics. Scopes, who openly admitted violating the law, was convicted, but his fine was rescinded by the state supreme court to avoid a legal technicality for testing the law's constitutionality.

ANNOTATION KEY

Answers

1 The hominids walked upright. The smaller footprints may be those of a child.

Integration

1 Life Science: Blood. See *Human Biology and Health*, Chapter 4.

3–2 (continued)

CONTENT DEVELOPMENT

The molecular structure of protein molecules can be used to show relationships between organisms. Because proteins are used to build and repair body parts, the more similar the structure of the protein molecules, the closer the relationship between organisms. Scientists have developed a scale that can be used to estimate the rate of changes in proteins over time. This scale is called a molecular clock.

In explaining the molecular clock, tell students that the DNA in an organism controls its production of proteins. As different species evolve from a common ancestor, their DNA becomes different. There will also be differences, therefore, in the proteins produced by the organisms. By examining the difference be-

Figure 3–9 *These 3.6-million-year-old footprints, which were fossilized in volcanic ash, are being unearthed by archaeologists in Laetoli, Tanzania. The trail on the left was made by three hominids. The prints on the right are those of a three-toed horse. What do the footprints tell you about these hominids?* 1

hands indicate she was adapted to life mainly in the trees. Tree life, of course, would disqualify her as the first human ancestor.

A few years after the discovery of Lucy, British anthropologist Mary Leakey found a trail of footprint fossils in Tanzania in Africa. The fossil footprints seem to have been made by two hominids, walking side by side. Perhaps the larger hominid, the parent, was holding the hand of the smaller one, its offspring. Whatever else these fossil footprints show, they make one thing clear—whoever made them walked upright on two legs, as humans do. And, when the mud in which the footprints were embedded was analyzed, it was found to be about 4 million years old. This evidence showed that these hominids walked the Earth 4 million years ago!

Over the years, more hominid fossils have been discovered and placed in the genus *Australopithecus*. To date, there are two other species in addition to *A. africanus* and *A. afarensis*. They are *A. robustus* and *A. boisei*. Which species is actually the earliest ancestor of humans? Might an even younger fossil qualify instead, or perhaps an older fossil? Could the true human ancestor still be hidden somewhere in the Earth's soil? The debate and the search continue.

Fossil Evidence Versus Chemical Evidence

Until about 1970 evidence for human evolution had come primarily from fossils. Based on this evidence, most scientists estimated that apes and humans began to take separate evolutionary paths more than 14 million years ago.

In the 1970s, new laboratory evidence emerged to contradict this concept. A new method for measuring differences between the proteins of different species had been developed. Scientists had also developed a scale that could be used to estimate the rate of change in proteins over time. As you may recall from Chapter 2, this scale of protein change was referred to as a molecular clock.

In one case scientists compared the protein structures of hemoglobin in various modern primates, including humans. Hemoglobin is the red pigment in blood. The hemoglobin of humans and

---

tween similar proteins in two related species, scientists can determine when they took separate paths in their evolution from a common ancestor.

When scientists compared the structure of human and chimpanzee hemoglobin molecules, they found that the molecules had the exact same sequence of 287 amino acids. (Amino acids are the building blocks of protein.) When the hemoglobin molecules of a human and a gorilla were compared, scientists

found that two of the amino acids had different positions. This led scientists to conclude that gorillas took a different evolutionary path before the separation of the paths taken by chimpanzees and humans.

• **Would you expect the hemoglobin molecules of a house cat and a tiger to be similar? Explain your answer.** (Students should suggest a close similarity of molecules based on the similar structure of the animals.)

chimpanzees had exactly the same sequence of 287 amino acids, which are the building blocks of proteins. The hemoglobin of humans and gorillas, however, differed in the position of 2 amino acids. This evidence led scientists to conclude that gorillas took a separate evolutionary path before the paths of chimpanzees and humans separated.

When did the gorilla path split off? According to the molecular clock, each change in the hemoglobin molecule would have taken 3 to 4 million years to occur. So gorillas would have taken a separate path 6 to 8 million years ago. Chimpanzees and humans would have taken separate paths more recently. This suggested split was much more recent than the 14-million-year date provided by fossil evidence. Today many scientists estimate that this split in evolutionary paths took place as recently as 2 million years ago.

Skillful Human: *Homo habilis*

The first species of hominid to actually be called human was also the maker of the first tools. *Homo habilis* (ha-BIH-lihs), which means skillful human, lived about 2 million years ago. The fossils of *Homo habilis* were first discovered in the 1960s in Olduvai Gorge in Kenya by the Leakey expedition. Olduvai Gorge is the oldest settlement of humans yet discovered.

In 1972, Kenyan anthropologist Louis Leakey (husband of Mary Leakey) reported the discovery of

Figure 3–10 *Based on the molecular clock, or scale of protein change, scientists determined that gorillas (right) took a separate evolutionary path before the paths of chimpanzees (left) and humans separated.*

F ■ 85

Anthropologists have pieced together evidence from fossils and relics to provide a picture of what daily life must have been like for *Homo habilis*, the earliest human ancestor.

The first humans may have lived in trees. A recently discovered fossil skeleton reveals that *Homo habilis* had long arms adapted for tree climbing. *Homo habilis* ate plant parts such as berries, roots, and tubers. They also scavenged meat from the kills of other animals. The young matured twice as fast as modern humans, needing much less parental care. Although lacking the language abilities of modern humans, *Homo habilis* had the neurological structures in their brains associated with speech as well as a larynx adapted to making a broad range of sounds.

marks fell on top of the tooth marks.
- **What does the evidence suggest about the order in which the two marks—tool and tooth—were placed on the bones?** (It suggests that the tooth marks were on the bones first, before the tool marks.)
- **What does this suggest about what killed the animals?** (It suggests that the large carnivores killed the animals because the carnivores would have killed with their teeth. Humans would have killed with tools.)

Point out that this conclusion could be supported by the fact that the tool marks on these bones were not where the most meat would be. Instead, they were in places where shreds of meat would be left over after a carnivore had eaten.

INDEPENDENT PRACTICE

▶ *Activity Book*

Students may enjoy hypothesizing about adaptations of physical characteristics that would have enabled early human ancestors to live in trees by completing the chapter activity Shipwrecked! In the activity students make inferences about the physical differences between modern humans and early ancestors of humans.

● ● ● ● **Integration** ● ● ● ●

Use the information about molecular clocks to integrate life science concepts of blood into your lesson.

GUIDED PRACTICE

Skills Development

Skills: Making inferences, drawing conclusions

Point out that when anthropologists first studied the bones of animals used

by *H. habilis* for food, they noticed that many of the animal bones showing marks from stone tools also showed tooth marks from large carnivores. At first, the anthropologists thought that the large carnivores must have scavenged from the leftovers after *H. habilis* had finished with the carcass. Later on, however, new evidence was obtained by examining the bones with a scanning electron microscope. The scanning electron microscope revealed that on many bones, the tool

BRAIN SIZE

Chimpanzees, our closest living relatives among the apes, have a brain size of about 280 to 450 cubic centimeters. The brain of *Homo sapiens,* however, ranges in size from 1200 to 1600 cubic centimeters. Most of the difference in brain size results from the enormously expanded cerebrum of *Homo sapiens.* The cerebrum is the "thinking area" of the brain.

Figure 3–11 *This skull of* Homo habilis, *which was found in Kenya, is about 1.8 million years old. The pebble tools, which are thought to be associated with* Homo habilis, *may have been used to prepare food.*

more fossils of *H. habilis.* These fossils included stone tools, such as the ones shown in Figure 3–11. These tools may have been used to prepare vegetables and to hunt smaller animals.

Upright Human: *Homo erectus*

The next humans known to scientists lived in caves, where they kept themselves safe from animals and from bad weather. These humans used fire to provide heat and light for their caves. Interestingly, however, these humans did not know how to build fires. So they waited until lightning set fire to a nearby bush, and then they carefully brought the fire back to their caves. Over the fire, they cooked the meat of animals and roasted nuts and seeds.

Along with fossils of these hominids, scientists have found carefully chipped hand axes. Painstaking examination of these tools has given scientists a good idea of how the tools were made. It may have happened something like this: Holding a small piece of sandstone in one hand, a hominid would strike the sandstone against a flat rock again and again, chipping off small pieces. These pieces were then gathered. If they happened to be the right size, they were rubbed into blades and points that were used to make weapons for hunting small animals.

This description is about all that is known of the life of the early human called *Homo erectus* (eh-REHK-tuhs). *H. erectus* lived from about 1.6 million to 500,000 years ago. *H. erectus* was the earliest human species to move out of Africa. Fossils of *H. erectus* have been found on the island of Java in Indonesia, near Heidelberg, Germany, and near Beijing, China. *H. erectus* had thicker bones than modern humans do, a sloping forehead, and a very large jaw. Such a large jaw would have been needed to chew some of the tough foods these humans ate. No one knows precisely what became of *H. erectus.*

3–2 (continued)

CONTENT DEVELOPMENT

Explain to students that scientists believe that *Homo habilis* evolved into another early human called *Homo erectus.* According to fossil evidence, *Homo erectus* was a cave dweller who lived from about 1.6 to 0.5 million years ago. *H. erectus* stood about 1.5 meters tall, had a sloping forehead, a large chinless jaw, and a brain that was a little larger than the brain of *H. habilis.* Their bones were thicker than those of modern humans.

Although they did not understand how to build fires, *H. erectus* learned to control those started by lightning. *H. erectus* was a toolmaker who used stone tools.

Point out that because *H. erectus* fossils were first found on the Southeastern Asian island of Java and in China, H. erectus is sometimes called Java Man and Peking Man.

GUIDED PRACTICE
Skills Development
Skills: Making inferences, drawing conclusions

Emphasize the fact that *H. erectus* could control fire but did not know how to build a fire.

• **Do you think H. erectus had fire only after a lightning storm that naturally produced a fire? Explain.** (Answers may vary. Some students may suggest that *H. erectus* kept fires burning at all times so that they would always be available.)

• **If H. erectus did keep fires burning at all times, how important would collecting kindling and tending the fire be to the species?** (Students should conclude that

Figure 3–12 Homo erectus *replaced* Homo habilis, *spreading throughout Europe, Africa, and Asia about 1 million years ago. Evidence strongly indicates that* Homo erectus *used and controlled fire.*

PROBLEM Solving

A Prehistoric Puzzle

Scientists have discovered fossils of *Homo erectus* in many places on Earth. In one site in Spain scientists were puzzled by a set of events. They found the remains of ancient brush fires on top of a cliff. At the base of the cliff they uncovered the bones of an entire herd of elephants. Scattered among the elephant bones were some stone tools.

Developing a Hypothesis
■ Suggest a hypothesis that would explain the events.
■ How can your hypothesis be used to explain the behavior of *Homo erectus*?

such jobs would be very important.)
• **What do you think *H. erectus* used fires for?** (Students might suggest that they used fires for warmth and for cooking.)
• **What might happen if the fires went out?** (Accept all logical answers. Students might suggest that *H. erectus* would have to wait until the next lightning storm started a fire, that they would have to eat raw foods, and that they might suffer from hypothermia.)

ACTIVITY
DISCOVERING

Communication

① Sign language is a system of hand signals that people use in order to communicate with others. Look up some words or phrases in sign language and show them to your class.
■ What are some other ways in which people communicate with one another?

Wise Humans: *Homo sapiens*

About 500,000 years ago a new species of hominids appeared on Earth. Because the fossils of this new species had a body skeleton much like modern humans, as well as a brain of similar capacity, the species was called *Homo sapiens* (SAY-pee-ehnz), or wise human. There are three groups of *H. sapiens*. One group, which is called early *H. sapiens*, lived about 500,000 to 200,000 years ago. Unfortunately, because fossils of this group are scarce, scientists know very little about early *H. sapiens*. As you will soon learn, however, there is more fossil evidence concerning the remaining two groups of *H. sapiens*.

NEANDERTHALS The period from about 150,000 years ago to about 35,000 years ago produced an abundance of fossils of what are now called **Neanderthals.** Neanderthals, who lived in parts of Africa, Asia, and Europe, received their name from the Neander Valley in Germany, where their fossils were first discovered. They were given the genus and species *Homo neanderthalensis* (nee-AN-der-thawl-ehn-sihs). The species name *neanderthalensis* remained with these hominids for many years. The name was changed, however, when more complete fossil evidence came to light. This evidence indicated that although Neanderthals were more heavily built than modern humans are, they stood as erect as modern humans do and had a brain capacity as large as that of modern humans. Neanderthals fished and hunted birds and large animals. They also used handmade stone tools. As a result of all this evidence, Neanderthals were placed in the same species as

Figure 3–13 *This illustration shows two of the many theories about the trend of human evolution. What are the major differences between these two theories?* ①

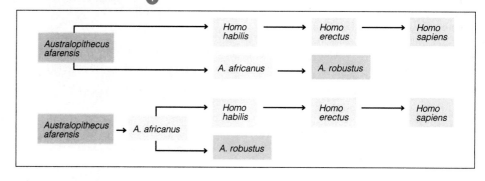

CONTENT DEVELOPMENT

Explain that scientists have identified three groups of *Homo sapiens,* but they have been able to provide information about only two of the groups. The Neanderthals lived in Europe and Asia from about 100,000 years ago to about 35,000 years ago. Although they walked on two

legs as modern humans do, they had heavy bones and a heavy bony eyebrow ridge. They were about 1.5 meters tall but were powerfully built. According to skull measurements, the brain size of Neanderthals was as large or slightly larger than that of modern humans, but the brain was shaped differently.

Stress to students that the most distinctive characteristic of the Neanderthals was that they buried their dead, sometimes with weapons and foods. This ritu-

might suggest that the objects are buried to show what was important to the culture.)

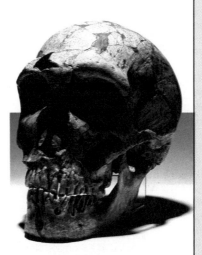

modern humans—*Homo sapiens*. But because Neanderthals were not truly identical to modern humans, the word *neanderthalensis* was added. So Neanderthals are now known as *Homo sapiens neanderthalensis*.

Although Neanderthals sometimes lived in caves, it is likely they moved from place to place in search of food. When Neanderthals camped out in the open, they probably built temporary huts by stretching animal skins over a framework of bones. The burnt wood found in Neanderthal camps suggests that they were experts at controlling fire. They may even have known how to start a fire with flint.

More impressive than the Neanderthals' ability to hunt, fish, cook, and make tools is the fact that they must have been the first hominids to act according to beliefs and feelings about the nature of the world. Like the ancient Egyptians, who lived many thousands of years later, Neanderthals buried their dead with the tools, herbs, or animal bones that had been important to the dead individuals in life. Sometimes the Neanderthals arranged animal bones around the graves in patterns that suggest religious rituals.

CRO-MAGNONS The first humans identical to modern humans began to appear on Earth about 40,000 years ago. These large-brained people were called **Cro-Magnons** (kroh-MAG-nuhnz), after the place in southwestern France where they were first discovered. However, similar fossils dating back 92,000 years ago have been found in Qafzeh, Israel.

Figure 3–14 *Neanderthals are the oldest known hominids to have buried their dead with objects that include tools, animal bones, and even herbs.*

Figure 3–15 *This Neanderthal skull, which was found in France, is between 35,000 and 53,000 years old. The pattern of wear on the teeth indicates that the hominid probably used its teeth for more than eating—perhaps for softening hides.*

F ■ 89

BACKGROUND INFORMATION
WHAT HAPPENED TO THE NEANDERTHAL?

There is ample fossil evidence that the Neanderthals lived side by side with the Cro-Magnons in several locations. Then about 30,000 years ago, the Neanderthals disappeared. Some scientists think that the Cro-Magnon interbred with the Neanderthals, blending their characteristics. Other scientists think that the Cro-Magnons slaughtered the Neanderthals. What caused the Neanderthals to disappear remains a mystery.

al suggests that the Neanderthals were the first early humans to have advanced beliefs or feelings about the natural world around them. Stress that although the Neanderthals were much more advanced than were previous human species, they were not a successful species and disappeared about 35,000 years ago.

Point out that Neanderthals were replaced by Cro-Magnons, a seemingly more advanced human species. Cro-Magnons had long faces, straight high foreheads, and small teeth. Point out that Cro-Magnons worked together to build shelters and tools and to hunt for food, and they probably used language to communicate.

● ● ● ● **Integration** ● ● ● ●

Use the comparison to Neanderthal and Egyptian burial practices to integrate social studies concepts into your lesson.

GUIDED PRACTICE

Skills Development
Skills: Applying concepts, hypothesizing, using nonverbal communication

Divide the class into several groups of four to six students. Tell the groups that they are to develop a system of nonverbal communication to explain to one another the performance of a specific task, such as caring for a sick person or asking someone to build a fire. Through this activity students should come to understand the significance of oral language and the importance that all members of a group understand the symbols or gestures that indicate meaningful communication.

• **Is oral communication easier to understand than nonverbal communication?** (Answers will vary. Most students will agree that verbal language is easier to understand.)

• **What happens if a member of a group cannot interpret the gestures or symbols of a communication?** (It is more difficult to communicate if a member does not understand the meaning of a message.)

ACTIVITY
WRITING
THE ICEMAN

Before students begin their research, explain that the *Readers' Guide to Periodical Literature* will identify magazine articles about the 4000-year-old man. Tell students that they might look under such headings as Iceman from Similaun, Similaun Glacier, artifacts, Bronze Age, or anthropology to find articles related to the story. The Iceman was remarkably well preserved because he was locked in a frozen crevasse in the Tyrolean Alps.

Integration: Use this Activity to integrate language arts skills into your lesson.

ANNOTATION KEY

Integration
① Language Arts
② Physical Science: Electromagnetic Radiation. See *Sound and Light,* Chapter 3.

3–2 (continued)

REINFORCEMENT/RETEACHING

▶ *Activity Book*

Students who need additional help in reinforcing concepts of human evolution can complete the chapter activity An Evolutionary Chart. In the activity students identify characteristics of different human ancestors.

GUIDED PRACTICE

▶ *Laboratory Manual*
Skills Development

Skills: Manipulative, making comparisons, drawing conclusions, making inferences

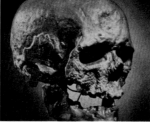

Figure 3–16 *Cro-Magnons, who were the first hominids truly identical to modern humans, appeared on Earth about 100,000 years ago. New and different tools, such as the laurel-leaf blade made from flint and used as the point of a spear, characterized their culture.*

ACTIVITY
WRITING

The Iceman

① Using reference materials in the library, find out about the 4000-year-old man discovered in September 1991 preserved in ice on the Similaun Glacier. This glacier is found in a pass between Austria and Italy. Describe the clothing that this prehistoric man wore and the significance of the bronze-headed ax he carried.

Cro-Magnons were thinner than the Neanderthals were and had a more complex culture. Cro-Magnons also made more advanced tools, such as spears, fishing hooks, and needles. And, Cro-Magnons produced some of the most creative works of art in early human history. Cro-Magnons are now placed in the same group as modern humans are—*Homo sapiens sapiens*—meaning the wisest of the wise.

Scientists have carefully studied evidence that might provide additional clues to Cro-Magnon life. Such evidence indicates that Cro-Magnons worked together to make tools, build shelters, and hunt. To do so, they probably spoke to one another. Because they did not leave written records, no one has any idea what their language may have been like.

Whatever the language may have been, its importance for the future development of humans right up to today cannot be underestimated. Language is used to spread ideas. Although many animals communicate by sounds, only humans have developed a complicated communication system capable of transferring much of what goes on in one person's mind to that of another.

Language can be thought of as a final step in the evolution of the modern *H. sapiens sapiens.* Largely because of language, humans have developed a new characteristic: the ability to describe, or examine, one's own existence. As far as anyone knows, humans are the only animals capable of talking about themselves and of peering back down the long path of history to their own beginnings.

3–2 Section Review

1. What are some adaptations of human ancestors?
2. What evolutionary paths led to modern humans?
3. Which genus was the first hominid to walk upright?

Critical Thinking—*Relating Facts*

4. Suppose you were able to meet a group that consisted of the following: *Homo erectus, Homo sapiens sapiens, Homo sapiens neanderthalensis,* Lucy, and *Australopithecus africanus.* How would you distinguish one from the other?

At this point you may want students to complete the Laboratory Investigation called A Human Adaptation found in the *Laboratory Manual.* In the investigation students determine whether there is a genetic variation in the amount and effectiveness of ptyalin in class members.

CONTENT DEVELOPMENT

Stress to students that language development was crucial to the future development of humans. Language is used to spread ideas and communicate thoughts and feelings from one person to another. Stress that most of the evolutionary changes in humans up to the time of the Cro-Magnons were physical changes. With the development of the Cro-Magnons, however, changes have been mainly behavioral and cultural. Much of this change can be attributed to the development of more complicated systems of communication.

CONNECTIONS

Dating a Neanderthal ❷

In early 1991, at an archaeological site near the village of St.-Césaire, north of Bordeaux, France, scientists discovered a Neanderthal skeleton. The fossil provided evidence that Neanderthals lived in Western Europe as recently as 36,000 years ago. This date is several thousand years after the first modern humans were believed to have appeared there.

How were the scientists able to determine the age of the fossil? Because the usual technique of radiocarbon dating is not effective with objects so old, scientists had to use another method of dating. The technique the scientists employed is called *thermoluminescence* (ther-moh-loo-muh-NEHS-ehns), and it was used on the flint tools found with the skeleton. Thermoluminescence involves the study of the *light energy* released when flint is heated to a temperature of about 450°C. By heating the flint to this temperature, the electrons (negatively charged particles) in the atoms of the flint are rearranged. Their changes in position release energy. The amount of energy released is analyzed to determine how much time has passed since the flint was last heated.

In this case, thermoluminescence showed that the age of the flint tools was about 36,300 years, give or take 2700 years. From this evidence, the scientists concluded that the bones of the Neanderthal skeleton were the same age as the flint tools. A fossil skeleton of the most recent Neanderthal yet known had been identified using energy in the forms of heat and light.

Students may be fascinated by the use of light energy for dating fossils. Point out that the dating procedure for determining the age of the Neanderthal skeleton was indirect. The tools of the Neanderthal were used to determine its age. Thermoluminescence can be used only on fossils that had been previously burned; the flint tools were therefore used to establish the age of the skeleton.

If you are teaching thematically, you may want to use the Connections feature to reinforce the themes of energy, evolution, and systems and interactions.

Integration: Use the Connections feature to integrate physical science concepts of electromagnetic radiation into your lesson.

would have the ability to communicate through a verbal language. *Homo sapiens neanderthalensis* would be heavily built and would have handmade stone tools. Lucy may have elongated arms and hands adapted to tree living. *Australopithecus africanus* would have apelike and humanlike characteristics.

REINFORCEMENT/RETEACHING

Monitor students' responses to the Section Review questions. If students appear to have difficulty with any of the concepts, review the appropriate material in the section.

CLOSURE

▶ *Review and Reinforcement Guide*

At this point have students complete Section 3–2 in the *Review and Reinforcement Guide.*

INDEPENDENT PRACTICE

Section Review 3–2

1. Human ancestors stood upright, which freed their hands for other tasks; had opposable thumbs, which enabled them to grasp objects easily; had relatively large and complex cerebellums as well as other evolutionary characteristics.

2. Many models for the evolutionary path exist. One model begins with *Australopithecus afarensis*, which evolves into two separate paths. One develops into *Homo* *sapiens* through *Homo habilis* and *Homo erectus*. The other evolves into *A. africanus* and then into *A. robustus*.

3. *Australopithecus.*

4. Answers will vary but should be based on the physical and behavioral characteristics of the species. Students might suggest that *Homo erectus* were cave dwellers who could not build fires and who had thicker bones than modern humans. *Homo sapiens sapiens* would most closely resemble modern humans and

Laboratory Investigation

COMPARING PRIMATES—FROM GORILLAS TO HUMANS

BEFORE THE LAB

At least one day prior to the investigation, gather enough materials for the class, assuming six students per group.

PRE-LAB DISCUSSION

Have students read the complete laboratory procedure.

- **What is the purpose of this laboratory investigation?** (To identify changes that have occurred among early hominids and hominids of today.)
- **Why do you measure your own jaw and thumb indexes?** (As an example of a modern human for comparison to early hominids.)
- **Do you think the illustrations are drawn to scale? Why?** (Most students will understand the illustrations are drawn to scale to ensure that the measurements are correct.)
- **Would you predict that your jaw index will be greater or smaller than that of *Australopithecus*? Why?** (Accept all predictions. Many students will predict that their jaw index will be smaller because earlier hominids had larger jaws.)

Laboratory Investigation

Comparing Primates—From Gorillas to Humans

Problem

What changes occurred as humans evolved from earlier primates?

Materials *(per student)*

scissors	metric ruler
clean paper	protractor

Procedure

1. Insert a 6 cm X 9 cm strip of paper lengthwise into your mouth. Place the paper over your tongue so that it covers all your teeth, including your back molars. Bite down hard enough to make an impression of your teeth on the paper. Remove the paper from your mouth.
2. Draw a line on the paper from the center of the impression of the left back molar to the center of the impression of the right back molar. Mark the midpoint of this line. Use the protractor to draw a perpendicular line from the midpoint of the line connecting the back molars to the front teeth.
3. Measure the width of the jaw by measuring the length of the line between the back molars. Measure the length of the jaw by measuring the line from the back of the mouth to the front teeth. Record your measurements in a data table.
4. Calculate the jaw index by multiplying the jaw width by 100 and then dividing by the length of the jaw. Record the jaw index.
5. Repeat steps 3 and 4 using the drawings of gorilla and *Australopithecus* jaws.
6. Find the indentation at the bottom of your palm near the ball of your thumb. Measure the length of your thumb from the indentation to its tip. Measure the length of

your index finger from the indentation to its tip. Record these measurements.
7. Calculate the thumb index by multiplying the thumb length by 100. Divide by the index-finger length. Record the thumb index.
8. Repeat steps 6 and 7 using the drawings of the thumb and index finger of both the gorilla and *Australopithecus*.

Observations

What trend did you observe regarding the relative length of the jaw? The relative length of the thumb and index finger?

Analysis and Conclusions

1. Was *Australopithecus* a mammal with characteristics somewhere in between those of gorillas and humans? Give evidence to support your answer.
2. Based on the thumb index, what adaptive change occurred in human evolution? What was the advantage of this change?
3. Based on your observations, what other change occurred as humans evolved?

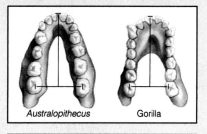

| *Australopithecus* | Gorilla |

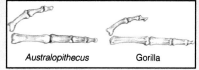

| *Australopithecus* | Gorilla |

TEACHING STRATEGY

1. Discuss with students how they might set up their data table. You might write a basic data table on the chalkboard for students to use as a model.
2. If necessary, show students how to draw the lines on their own impressions for making measurements.
3. To help students with the mathematics, write the needed equations on the chalkboard and demonstrate some sample problems. Allow students to use calculators.

DISCOVERY STRATEGIES

Discuss how the investigation relates to the chapter ideas by asking open questions similar to the following.

- **Would you expect the gorilla or *Australopithecus* to have greater dexterity?** Why? (*Australopithecus* probably would have greater dexterity because its thumb was longer—making inferences, hypothesizing.)
- **Do you think *Australopithecus* was a closer relative to modern humans than the gorilla is?** (Based on their investiga-

Study Guide

Summarizing Key Concepts

3–1 The Search for Human Ancestors

▲ Primates are members of a group of mammals that includes humans, monkeys, and about 200 other species of living things.

▲ Primates share several important characteristics: They have flexible hands with opposable thumbs; they can see in three dimensions; and they have large, complex cerebrums.

▲ Some 50 million years ago, early primates split into two main evolutionary groups: prosimians and anthropoids.

▲ The anthropoids then split into two branches. One branch evolved into New World monkeys. The other branch evolved into Old World monkeys and hominoids; hominoids include gorillas, gibbons, orangutans, chimpanzees, and humans.

3–2 Human Ancestors and Relatives

▲ About 6 million years ago, hominoids gave rise to a small group of species called hominids, which include humans and closely related primates.

▲ Early hominids experienced changes in the shapes of their spinal column and their hip and leg bones. These changes enabled them to walk upright and thus freed their hands to use tools. At the same time the opposable thumb evolved.

▲ The first recognized hominids were called *Australopithecus*.

▲ Chemical and fossil evidence indicates that the split between apes and humans may have occurred 2 million years ago.

▲ The first species to be placed in the genus *Homo* was skillful human, or *Homo habilis*.

▲ *Homo habilis* was followed by *Homo erectus*.

▲ The first species to resemble modern humans seems to have evolved 150,000 years ago and is called *Homo sapiens neanderthalensis*.

▲ The first fossils of modern humans date to about 100,000 years ago. These humans, called Cro-Magnons, are in the genus *Homo sapiens sapiens,* as are modern humans.

Reviewing Key Terms

Define each term in a complete sentence.

3–1 The Search for Human Ancestors
primate

3–2 Human Ancestors and Relatives
Neanderthal
Cro-Magnon

Part 1

Have students conduct outside research to find out how tooth size has changed throughout human evolution. Then have them use the information to design an investigation in which they compare tooth sizes of modern humans to those of several earlier species.

An interesting part of such an investigation would be observing the different-sized teeth among members of the class. Scientists believe that variations among humans today are part of a trend toward smaller teeth—meaning that human teeth are still changing.

Part 2

Suggest that students find out the jaw index of species from different groups. For example, they might find the jaw indexes of a house cat and a lion. Ask students to draw conclusions about the different species based on the indexes.

---

tions, students might conclude that *Australopithecus* is a closer relative because the physical features studied are closer to that of modern humans—drawing conclusions.)

• **Would the data differ if a baby gorilla's measurements were compared to that of an adult human?** (The results would be similar because the measurements would be proportionate to size—making inferences, applying concepts.)

OBSERVATIONS

The jaw became progressively shorter in going from *Australopithecus* to humans. However, the thumb became progressively longer.

ANALYSIS AND CONCLUSIONS

1. The jaw index and thumb index are intermediate to those of the gorilla and human.

2. The thumb became relatively longer, making it more able to reach across the hand and grasp objects.

3. The jaw became shorter, producing a more V-shaped dental arch.

Chapter Review

Chapter Review

ALTERNATIVE ASSESSMENT

The *Prentice Hall Science* program includes a variety of testing components and methodologies. Aside from the Chapter Review questions, you may opt to use the Chapter Test or the Computer Test Bank Test in your *Test Book* for assessment of important facts and concepts. In addition, Performance-Based Tests are included in your *Test Book*. These Performance-Based Tests are designed to test science process skills, rather than factual content recall. Since they are not content dependent, Performance-Based Tests can be distributed after students complete a chapter or after they complete the entire textbook.

CONTENT REVIEW

Multiple Choice

1. d
2. c
3. c
4. a
5. c
6. c
7. c
8. d

True or False

1. F, Primates
2. F, Old World
3. F, Raymond Dart
4. F, 3.5
5. F, skillful
6. T
7. T

Concept Mapping

Row 1: Anthropoids
Row 2: Old World Monkeys
Row 3: Hominoids

CONCEPT MASTERY

1. Primates have forward-pointing eyes, three-dimensional vision, flexible hands, large and complex cerebrums; most have opposable thumbs. Humans walk on two legs instead of four, have a wider pelvis, and communicate by language.
2. The genes and cultures of the Neanderthals and Cro-Magnons could have blended and could have been passed on to modern humans.
3. Scientists compare the protein structures of various species to determine whether there are any differences. If the protein structure is exactly alike, the species are closely related. If the protein structure is different, they are not closely related. In this case, scientists measure the differences and estimate the rate of change in proteins over time. This is referred to as the molecular clock.
4. *H. habilis* lived about 2 million years ago. Fossils indicate that *H. habilis* had hands capable of performing delicate tasks. They also are believed to have stood upright. *H. erectus* lived about 1.6 to 0.5 million years ago. They lived in caves, used fire, and made tools. Fossils of *H. erectus* indicate that they had thicker bones than do humans, a sloped forehead, and a very large jaw.
5. From fossil evidence *A. afarensis* walked on two legs and was adapted to life mainly in trees. *A. africanus* had small teeth, signs of upright posture, and a brain whose size lies somewhere between apes and humans.
6. Neanderthals were more heavily built than were Cro-Magnons, who were thinner and had a more complex culture.

Content Review

Multiple Choice

Choose the letter of the answer that best completes each statement.

1. Which does not include humans, apes, and monkeys?
   a. mammals   c. primates
   b. vertebrates   d. *Homo sapiens*
2. Which is not a characteristic of all primates?
   a. three-dimensional vision
   b. flexible hands
   c. two-legged walking
   d. larger cerebrum
3. What characteristic do humans have that other primates do not?
   a. sharp vision
   b. three-dimensional vision
   c. two-legged walking
   d. opposable thumbs
4. The molecular clock is used to study changes in
   a. proteins.   c. muscle tissue.
   b. bones.   d. brain size.

5. The humans who first skillfully made tools were
   a. *Australopithecus africanus.*
   b. *Homo sapiens.*
   c. *Homo habilis.*
   d. *Homo erectus.*
6. The first humans known to bury their dead were
   a. *Homo habilis.*   c. Neanderthals.
   b. *Homo erectus.*   d. Cro-Magnons.
7. Humans that could control fire were
   a. Neanderthals.
   b. *Homo habilis.*
   c. *Homo erectus.*
   d. *Australopithecus africanus.*
8. The first humans believed to have used language were
   a. *Homo habilis.*   c. *Homo erectus.*
   b. *A. africanus.*   d. Cro-Magnons.

True or False

If the statement is true, write "true." If it is false, change the underlined word or words to make the statement true.

1. <u>Hominids</u> are a group of mammals that includes humans, apes, and monkeys.
2. Gibbons and chimpanzees are examples of <u>New World</u> monkeys.
3. <u>Louis Leakey</u> discovered the first fossil of a hominid and named it *Australopithecus.*
4. The humanlike fossil nicknamed Lucy is about <u>14</u> million years old.
5. *Homo habilis* means <u>upright</u> human.
6. Primitive huts have been found in <u>Neanderthal</u> campsites.
7. <u>Cro-Magnon</u> is included in the species *Homo sapiens sapiens.*

Concept Mapping

Complete the following concept map for Section 3–1. Refer to pages F6–F7 to construct a concept map for the entire chapter.

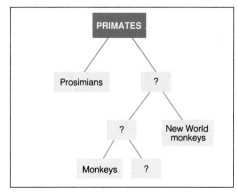

Concept Mastery

Discuss each of the following in a brief paragraph.

1. Describe the characteristics of primates. Explain how humans are different from other primates.
2. Explain how a blending of Neanderthals and Cro-Magnons could have led to modern humans.
3. Explain how protein structures are used to measure differences between species. Include a definition of molecular clock in your explanation.
4. Compare *H. habilis* and *H. erectus*.
5. Compare *Australopithecus afarensis* with *Australopithecus africanus*.
6. Explain how Neanderthals and Cro-Magnons differ.
7. Discuss the benefits that result from the ability to use language.
8. Compare anthropoids, hominoids, and hominids.
9. Discuss possible reasons for the disappearance of the Neanderthals.

Critical Thinking and Problem Solving

Use the skills you have developed in this chapter to complete each of the following.

1. **Interpreting graphs** Another characteristic that distinguishes humans from other primates is the length of time parents care for their young. The graph illustrates the length of time needed for three different species of primates to reach adulthood. Use the graph to answer the following questions.
   a. Which organism has the shortest preadult stage? The longest?
   b. How is the length of the preadult stage related to the intelligence of each primate species?

2. **Relating concepts** People often say that evolution means that humans evolved from monkeys and apes. Explain why such a statement is not an accurate representation of human evolution.
3. **Relating facts** Explain how the characteristics of primates made them successful at living in trees.
4. **Making comparisons** What advantages does a primate with the ability to walk on two legs have over a primate that walks on four legs?
5. **Relating concepts** How is the ability to learn language related to the development of human civilizations?
6. **Applying concepts** Do you think evolution on Earth has stopped? If not, do you think evolution will ever stop? Explain your answers.
7. **Using the writing process** With the exception of modern humans after the time of Cro-Magnon, choose a hominid and write a short story entitled "A Day in the Life of a Hominid."

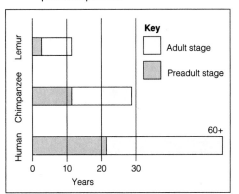

Key
☐ Adult stage
▨ Preadult stage

---

which enables them to perceive depth and judge distances when climbing trees and moving from one tree to another.

4. The ability to walk on two legs enables a primate to free its hands for different tasks. A primate that must walk on all four legs is not able to free its hands.

5. By learning language, humans spread their ideas and are able to describe their existence to future generations.

6. Answers will vary, but most students should respond that evolution is an ongoing process.

7. Students' short stories should reflect and be consistent with the scientific understandings about the hominid featured in the stories.

KEEPING A PORTFOLIO

You might want to assign some of the Concept Mastery and Critical Thinking and Problem Solving questions as homework and have students include their responses to unassigned questions in their portfolio. Students should be encouraged to include both the question and the answer in their portfolio.

ISSUES IN SCIENCE

The following can be used as a springboard for discussion or given as a writing assignment.

Have students debate the following statement: "Humans are the final product of evolution."

It is believed that Cro-Magnons had the ability to use language.

7. Language is used to spread ideas and to examine and describe one's own existence.

8. Anthropoids are an evolutionary group of primates that include monkeys, apes, and hominoids. Hominoids are apes and humans. Hominids are the closest relatives of humans.

9. Students might suggest that disease or evolutionary changes led to the disappearance of the Neanderthals.

CRITICAL THINKING AND PROBLEM SOLVING

1. (a) Lemurs, humans; (b) The longer the preadult stage, the more intelligent the primate is.

2. Humans did not evolve from other present-day primates. Rather, modern primates, including humans, have common primate ancestors.

3. Primates have flexible hands and opposable thumbs that enable them to grasp objects such as tree trunks and branches. They also have three-dimensional vision,

JACK HORNER
WARMS UP TO DINOSAURS

Background Information

In 1990, John R. Horner (Jack Horner) supervised the opening of the Phyllis B. Berger Dinosaur Hall in the Museum of the Rockies in Bozeman, Montana. The lifelike displays show dinosaurs as they may have appeared when they roamed the Earth. In one display a mother maiasaur feeds her nested young. It now appears that maiasaur and other dinosaurs nested and nurtured their young.

Also on view in the paleontology lab of the museum is the most complete skeleton of *Tyrannosaurus rex* ever found. The skeleton is 85 to 90 percent complete. The other skeletons of *T. rex* that were found were only about 50 percent complete.

The museum and its curator are changing people's understanding of the dinosaurs. They may not have been the coldblooded killers they have been portrayed to be. *T. rex* may have followed herds of *Triceratops,* scavenging carcasses and perhaps preying on the weak. It is unlikely, according to Horner, that *T. rex* and *Triceratops* engaged in fierce battles. At the museum, dinosaur bones are displayed in much the same way as they were excavated. Viewers can also watch lab workers clean and examine dinosaur fossils. The aim of the museum is to show dinosaurs as they lived and to show the work of paleontologists who try to understand the way of life of these ancient creatures.

GAZETTE

Jack Horner Warms Up To Dinosaurs

Paleontologist Jack Horner is not interested in how dinosaurs died. Rather, he is interested in how they lived.

Horner, who teaches at Montana State University and is also curator of paleontology at the Museum of the Rockies in Bozeman, Montana, heads the largest dinosaur research team in the United States. His work focuses on the behavior of dinosaurs, and his research findings are considered quite revolutionary. You see, Horner has discovered that these extinct creatures were not always the enormous, bloodthirsty monsters portrayed in horror movies and popular cartoons. More likely, Horner says, many dinosaurs were social plant-eaters and nurturing parents.

Horner's conclusions are based largely on research he has done in an area called the Willow Creek Anticline in his home state of Montana. He worked there for many years, uncovering fossils that date back to the Cretaceous Period, some 65 to 135 million years ago.

In studying the dinosaurs, Horner is ful-

filling a goal he has had since he discovered his first fossil at age eight. Throughout high school, he knew he wanted to be a paleontologist. But his schoolwork was rather poor and remained so throughout his college experience. (He flunked out of college seven times.) However, he continued taking and excelling in courses in paleontology, eventually earning an assistantship in paleontology at Princeton University. At age 31, while at Princeton, Horner discovered his academic problems were the result of dyslexia. Dyslexia is a condition in which the ability to read is impaired.

While still at Princeton, Horner became particularly fascinated with the study of juvenile dinosaurs—an area of paleontology that had received little or no attention. His fascination took him back to Montana where, quite by accident, he stumbled upon a collection of dinosaur egg fossils. The events leading to this discovery occurred something like this: While traveling through his native state in the late 1970s, Horner and a friend decided to stop for a brief rest at a small shop where rocks were sold. The owner, an

---

TEACHING STRATEGY:
ADVENTURE

FOCUS/MOTIVATION

Begin by displaying pictures of a number of animals tending their young. Be sure to include pictures of humans as well as birds, cats, dogs, and other familiar animals.

• **What do these pictures have in common?** (All show animals caring for their young.)

• **Do all animals care for their young?** (No. Some animals—for example, turtles—lay eggs and leave the eggs and hatchlings to survive on their own.)

• **Do you think dinosaurs cared for their young?** (Accept all answers.)

CONTENT DEVELOPMENT

Point out that Horner hoped to be a paleontologist from the time he found his first dinosaur bone. It did not seem

amateur fossil collector, asked Horner to identify some bones she had found. Upon close examination, Horner was amazed to see that the collection included some bone fragments from baby hadrosaurs, or duck-bill dinosaurs. No sooner had the shopowner showed Horner where she had found the fossils, than he began to dig.

Working at the Willow Creek Anticline through the 1980s, Horner and his team unearthed many dinosaur eggs and skeletons of young dinosaurs, particularly those of hadrosaurs. The shape and texture of the egg shells, as well as the structure of the baby dinosaur skulls, lead Horner to some new and exciting conclusions about the behavior of hadrosaurs. Baby hadrosaurs, Horner says, kept their babylike features—large heads with big eyes and shortened snouts—throughout their lives. These features encouraged adult dinosaurs to nurture them. Adult hadrosaurs, Horner believes, guarded their eggs before they hatched and then fed and protected their young after they were born. He has even given a name to the genus of dinosaur he discovered with these nurturing traits. He calls the genus *Maiasaura*, meaning good mother dinosaur.

Horner brings the past into view of the present and uses the latest techniques of the present to study the past. As curator of paleontology at the Museum of the Rockies, Horner wants his visitors to see how he believes dinosaurs lived during their 140-million-year reign on Earth. For example, he designs his displays to show dinosaurs doing such ordinary actions as scratching their jaws and sitting calmly on the ground. Horner has also coauthored a book on—what else?—dinosaurs. In addition, he is a consultant for a Japanese amusement park that

▲ A duckbilled adult oversees a nest of newly hatched maiasaurs. The hatchlings are tended to for months by adults.

has a huge display—complete with robot models—of the history of Earth.

Horner, the former college dropout, is now well into his forties. The years of struggling with dyslexia have not prevented him from earning the respect of fellow scientists, as well as the prestigious MacArthur Foundation fellowship. Interestingly, this award is also known as the "genius award."

GAZETTE ■ 97

Additional Questions and Topic Suggestions

1. Jack Horner's school career was hampered by an unidentified disability—dyslexia. Research information about dyslexia, what it is, and how it can be overcome. Find out about other well-known people who have overcome this disability.
2. Read Jack Horner's book *Digging Dinosaurs*. Prepare a book report in writing or by using graphics. Based on the book, tell what evidence scientists use to hypothesize about dinosaur life styles.
3. Find out more about the maiasaur and write a paper describing the day in the life of a baby maiasaur.

Critical Thinking Questions

1. Why might it be considered ironic that Jack Horner received a MacArthur Foundation fellowship? (He was a poor student who barely made it through college.)
2. How is the dinosaur exhibit at the Museum of the Rockies different from those at other museums? (It shows dinosaur bones as they were when dug up. It shows dinosaurs as they may have lived.)
3. Why do you think Jack Horner is more interested in how the dinosaur lived than in how the dinosaurs became extinct?
4. Horner believes that the dinosaurs may have been warmblooded animals. After researching information about the dinosaurs, write a paper supporting or disputing this belief.

likely that Horner would achieve this goal because of his dismal school record. Yet Horner persevered and achieved his goal despite a disability that went undiagnosed until he was an adult.
• **What lesson can be learned from Horner's story?** (Students might suggest that hard work and drive will help a person achieve his or her goals.)
• **How is Horner's work changing how people view the dinosaurs?** (Accept all logical answers. Lead students to understand that Horner is helping people understand a side of dinosaurs not generally portrayed in the media. He is showing them as nurturing parents.)

INDEPENDENT PRACTICE

▶ *Activity Book*

After students have read the Science Gazette article, you may want to hand out the reading skills worksheet based on the article in your *Activity Book*.

ISSUES IN SCIENCE

FROM RULERS TO RUINS: THE DEATH OF THE DINOSAURS

Background Information

Many theories have tried to explain the extinction of dinosaurs. One major theory proposes that the Earth's weather became so cold that the dinosaurs could not mate. As a result, the species became extinct. This theory is based on the geological record of ancient ice ages. During the Triassic Period, just over 200 million years ago, dinosaurs became the dominant species on Earth. Then about 65 million years ago, during the Cretaceous Period, dinosaurs became extinct. It is significant that this was just about when the Earth's climate began to cool.

Another theory claims that the dinosaurs contracted diseases caused by plant life that developed during the Cretaceous Period. These poisonous plants are not eaten by animals of today, but this theory contends that the dinosaurs ate the plants, became ill, and died. Yet another theory claims that a great epidemic spread across the land and killed the dinosaurs.

A new theory proposed by Luis and Walter Alvarez, a father-and-son team from California, suggests that a giant asteroid struck the Earth about 65 million years ago. The impact of the asteroid

FROM RULERS TO RUINS:
The Death of the Dinosaurs

They were lords of the Earth for over 135 million years—giant reptiles that thrived in the mild climate of prehistoric Earth, with its warm waters and lush foliage. Then suddenly, 65 million years ago, these unchallenged rulers of the planet disappeared completely. The great dinosaurs became extinct!

Imagine what the last days of the dinosaurs might have been like. . . . A *Triceratops* munches peacefully on a clump of green bushes. Nearby a baby *Stegosaurus* runs frantically from the deadly jaws of the *Tyrannosaurus*, king of the flesh-eating dinosaurs. A long-necked *Brontosaurus* wades into a pond as a *Pterodactyl* flies overhead. None of these animals notices the dark shadow that begins to move across the land. The shadow grows larger as a huge rock, the size of a mountain, blocks the sun.

The rock is hurtling toward Earth. When it strikes, a deafening boom and tremendous earthquakes shake the land. The heat generated by the collision starts fires that rage through the tropical forests. At the same time, the violent impact sends a mushroom-shaped cloud of dust into the atmosphere. The dust combines with the smoke and ash from the forest fires to form a thick blanket that spreads across the globe and blots out the sun.

For several months or longer, the huge dust cloud hovers in the atmosphere, preventing sunlight from reaching Earth. The

Earth becomes a dark, cold planet. Plants, which rely on sunlight to make their food, begin to die off. Plant-eating dinosaurs now have nothing to eat, and they too begin to die. And flesh-eating dinosaurs can no longer find nourishment. In all, almost 96 percent of all plant and animal species are killed during this time period!

This theory to explain the extinction of the dinosaurs is known as the asteroid-impact theory. Luis and Walter Alvarez, the father-and-son team from California responsible for the theory, have suggested that a large asteroid, perhaps 10 kilometers in diameter, struck Earth about 65 million years ago and began the chain of events you have just read.

The Alvarezes have evidence to support their theory. They have studied the layers of clay formed during the time of the dinosaur extinction. In the clay, they have found high levels of iridium, an element that is extremely rare on Earth. Their readings show levels of iridium 160 times greater than normal. Where could the iridium have come from?

The Alvarezes believe asteroids from outer space are the source. Asteroids are known to contain high levels of iridium. According to the asteroid-impact theory, iridium was deposited in the clay when the asteroid struck Earth and set off the series of events that killed the dinosaurs.

Although the evidence of iridium in the clay seems convincing, not all scientists agree with the theory. Critics say that the high levels of iridium can be traced to volcanic

---

TEACHING STRATEGY: ISSUE

FOCUS/MOTIVATION

Begin by asking students the following questions.

• **Suppose that a very common group of animals—like squirrels—was to become extinct. What might cause this to happen?** (Accept all answers. Possible answers include the following: Perhaps the squir-

rels contracted a disease; perhaps pollution or hunters killed the squirrels; perhaps the squirrels were killed by other animals; and perhaps the squirrels' food supply disappeared.)

• **Would the extinction of an animal such as the squirrel surprise you? Why or why not?** (Accept all answers. Many students will probably answer yes, saying that it is hard to imagine why an animal that is well adapted and prevalent would suddenly die out.)

CONTENT DEVELOPMENT

Continue the previous discussion by pointing out that dinosaurs were once a common sight on Earth. Not only were dinosaurs present in great numbers, but they were also considered rulers of the Earth—the very greatest animals of their time.

Emphasize that the extinction of such a great and populous group of animals must have been brought about by a significant change. The theories proposed

caused tremendous earthquakes, violent forest fires, and a thick cloud of dust and smoke in the atmosphere. As a result, the dinosaurs and almost 96 percent of all other living things on Earth perished. The Alvarez theory is supported by the fact that high iridium levels have been recorded in various impact craters. Iridium is rare on Earth, but asteroids are known to contain high levels of it.

Additional Questions and Topic Suggestions

1. Go to the library and read about other major animals that are extinct. Find out what conditions may have caused these animals to die out and discuss any controversy that may exist as to the cause for these animals' extinction.

2. What might be the significance of findings that the level of oxygen in the Earth's atmosphere was much higher during the age of dinosaurs than it was in the periods following their extinction? (Accept all logical answers. It may mean that dinosaurs required a higher concentration of oxygen than later animals; thus, a reduced supply of oxygen may have contributed to the dinosaurs' extinction.)

3. What fact do scientists agree on concerning dinosaur extinction? (There was a mass extinction of dinosaurs about 65 million years ago.)

4. What does this article illustrate about the relationship among individual species, planet Earth, the Milky Way, and the universe? (Accept all logical answers. Everything in the universe may be affected by an event somewhere else in the universe.)

by scientists have attempted to determine this change. Stress that it is difficult to determine which theory is correct because no humans were present when dinosaurs were alive.

REINFORCEMENT/RETEACHING

Have students list the various factors in the Earth's history that may have contributed to the extinction of dinosaurs. These factors include a significant cooling of the Earth's climate; the appearance of poisonous plants; evidence that supports the possible crash of an asteroid; the development of small furry animals; a significant change in the level of oxygen in the Earth's atmosphere; changes in sea levels; and possible evidence of a worldwide disease.

Discuss how each of these factors has contributed to a particular theory of dinosaur extinction. For example, evidence of significant cooling of the Earth lends credence to the theory that dinosaurs could not mate because the Earth's weather became too cold.

Class Debate

Have students debate this issue by arguing pro or con: "The extinction of the dinosaur was caused by an asteroid hitting the Earth, producing large clouds of smoke and dust that blocked out the sun."

eruptions, which bring iridium buried deep within the Earth to the surface and release it into the atmosphere.

But further study of the clay and sediment layers from the time of the dinosaur extinction continues to yield evidence supporting the asteroid-impact theory. Geologists have found quartz from the ancient sediment that contains cracks that could be the result of a single huge impact—such as the impact of an asteroid. Also, chemists have discovered in the clay a form of an amino acid that is almost nonexistent on Earth. It is, however, common in meteors.

Still, the debate continues. Some scientists protest that the extinction of the dinosaurs did not happen all of a sudden. They believe that the dinosaurs died out gradually due to changes in climate and sea level. The scientists support their claims with fossil evidence. Paleontologists have found dinosaur bones and eggs in sediment and clay layers formed nearly 64 million years ago. This is one million years after the asteroid-impact theory claims all the dinosaurs should have become extinct.

Gradual extinction of the dinosaurs could have been the result of major climate changes on the Earth over several million years. Many scientists think the Earth's temperature cooled dramatically around the time of dinosaur extinction. The resulting death of

plant life in turn led to the gradual downfall of the dinosaurs.

Still other scientists believe that the development of small furry mammals might have contributed to the dinosaurs' end. Tiny clever rodents that were able to outrun and outsmart the huge dinosaurs could have eaten the dinosaurs' eggs. As time passed, older dinosaurs would die and fewer young dinosaurs would be born, leading to the dinosaurs' extinction.

Recent findings have shown that the level of oxygen in the Earth's atmosphere during the time of the dinosaurs was much higher than it was in the periods following their extinction. Although this clue must still be studied in more depth before any conclusions are drawn, scientists are hopeful that this discovery will shed new light on the question of dinosaur extinction.

The factors that caused the downfall of the great dinosaurs are quite complex. Did an asteroid falling from space kill off the giant reptiles? Was it a change in climate or perhaps even a worldwide disease that destroyed the largest land creatures ever to walk the Earth? Could it have been tiny furry mammals that spelled doom for the dinosaurs? Perhaps a combination of all these factors brought about dinosaur extinction. What do you think?

ISSUE (continued)

GUIDED PRACTICE

Skills Development

Skill: Developing a model

Have students work in small groups. Ask each group to research the appearance and general dimensions of various dinosaurs. Then challenge each group to make chalk drawings showing the actual size of each kind of dinosaur. The drawings could be made in an outdoor area such as the school parking lot or playground.

GUIDED PRACTICE

Skills Development

Skill: Relating concepts

Discuss how the disruption of food chains may have contributed to the dinosaurs' extinction. For example, di-

nosaurs depended on plants for food; a change in climate may have caused the death of these plants. The impact of an asteroid may have caused a huge dust cloud, preventing sunlight from reaching the Earth; without sunlight, plants cannot make food. Once again, the death of plants may have deprived the dinosaurs of their food supply.

Emphasize that one of the reasons environmentalists are so concerned about preserving natural balances is that they

fear the disruption of food chains. Disruption in a food chain can cause the overpopulation of certain organisms and the possible extinction of other organisms.

INDEPENDENT PRACTICE

▶ *Activity Book*

After students have read the Science Gazette article, you may want to hand out the reading skills worksheet based on the article in your *Activity Book*.

EVOLUTION
ON VIVARIUM

The rules of evolution, if followed elsewhere in space, may have produced living things undreamed of on Earth.

FROM: **Dr. Toshi Kanamoto, Bio-Ship 80**
TO: **Dr. Peter Harrington, Command Station 40, Earth**
RE: **Life on Planet Vivarium**

Greetings Pete:

We are now orbiting the planet Vivarium and are T minus 3 hours 20 minutes to landing. In just 3 more hours, if all goes well, we should be the first Earthlings to touch down on Vivarium.

You can imagine the excitement I feel. According to the data we received from the last exploratory satellite, the atmosphere of Vivarium is rich in oxygen. In fact, there is more than enough oxygen to sustain life as we know it on Earth.

Dr. Susan Haley, our ship's anthropologist, and Captain Jasper Fernandez, our commander, are angry with me. I just showed them the photograph I had been holding of the ET, or extraterrestrial life form. The photograph was taken on Vivarium by the satellite we sent out six months ago. Both

---

FUTURES
IN SCIENCE

EVOLUTION ON VIVARIUM

Background Information

This intriguing article offers a look at the kind of life forms that may exist elsewhere in the universe. The basis for this speculation is the idea that living things develop in response to their environment; that is, they tend to develop characteristics that will enable them to adapt to such variables as light, heat, water or lack of water, and the like.

By noting the way life forms have evolved on Earth, some scientists feel that they can predict how life forms may have evolved on distant planets. An interesting aspect of this line of reasoning is the effect of gravity on the development of living things. The size, proportions, and structure of various animals must of necessity be mechanically practical. A large creature with spindly legs will fall down because the legs cannot support the mass of the body. Of course, if gravity is dramatically reduced—say to one sixth or one tenth of that of the Earth—then body design can be more daring because even a massive animal will not weigh very much. Reduced gravity would also mean that animals would not have to expend as much energy in moving around. Thus, the lungs and heart would be less developed.

As would be expected, gravity greater than that of the Earth would have the opposite effect on the physical development of life forms. One scientist points out that the bipedal structure of humans would be a colossal failure if the Earth's gravity were to be doubled. This scientist declares that under double-gravity conditions, most of the Earth's animals would resemble short-legged saurians or serpents. Even birds would have a problem surviving, although their problem would be partly compensated by the increased density of air.

TEACHING STRATEGY: FUTURE

FOCUS/MOTIVATION

Have students observe the pictures of extraterrestrial life forms on pages 102–103.

• **Do these look to you like living creatures?** (Accept all answers.)
• **Do they resemble any animals or plants you have seen on Earth?** (Accept all answers.)
• **What do you think might cause a living organism to develop in this way?** (Accept all answers.)

Additional Questions and Topic Suggestions

1. Scientists are able to predict possible sizes of animals according to the Square-Cube Law. What is this mathematical principle? Who first discovered it? (Discovered by Galileo, this principle states that as the size of an object increases, the volume always increases faster than the surface area.)

2. Make a drawing to show the kinds of life forms that you think Dr. Kanamoto will discover on Vivarium.

3. Get together with several classmates. Create a planet that has whatever environmental characteristics you wish. Then, using the principles outlined in this article as a guide, decide what type of life forms might evolve on your planet.

of Vivarium's atmosphere. Haley argued that the slits were sense organs, probably eyes. She contended that eyes would be an important adaptation to the rocky, lake-filled terrain of Vivarium.

I suggested that we consider some other senses that might be adaptations on Vivarium. Haley and Fernandez looked at me curiously. Finally Fernandez said, "Go on."

"Well, we know that each hemisphere of Vivarium is in semi-darkness part of each year," I reasoned. "Could the creature have evolved a way of seeing by means of heat waves, rather than light waves, to help it live in the dark?"

Haley and Fernandez were impressed with my hypothesis and urged me to continue.

"Let's not forget about the radioactive ores our satellite discovered on Vivarium. Maybe the creature has evolved a sense that is like a biological Geiger counter. This sense would warn the creature whenever it came near dangerous concentrations of radioactivity."

"Those are two strong possibilities," said Haley. She began scanning the photograph closely. "I wonder how big the ET is. If this is only the head, the entire creature may be three times this size."

Fernandez and I tripled the head proportions, took the scale of the photograph into account, and figured that the creature was more than 200 times larger than any insect on Earth. But Haley insisted that our calculations had to be wrong. She wanted to know how such a gigantic creature could possibly support its own mass. It would be

feel it was wrong of me to keep the photograph a secret until now. But I knew the photograph would create a storm of controversy among the three of us. I just wanted to avoid spending ten days in space arguing over what the photograph tells us. But I'm afraid Haley and Fernandez disagree with my reasoning.

The photograph of the ET shows what looks like the head of a land creature, at the edge of a lake. The head is shaped something like that of an ant. Fernandez, Haley, and I immediately agreed that its black, smooth "skin" is probably an exoskeleton, or outer skeleton. This would be similar to the exoskeletons of some of Earth's insects.

Haley and Fernandez next began to discuss the three slits on the surface of the creature's head, above the mouth. Fernandez insisted that all three openings were breathing holes. They would allow the creature to take advantage of the rich oxygen content

---

FUTURE (continued)

CONTENT DEVELOPMENT

Emphasize to students that life forms tend to develop in response to their environment; that is, they tend to develop characteristics that will enable them to adapt to the conditions in which they must live. Some scientists who have studied the evolution of life on Earth have begun to speculate about what types of life might develop on a planet different from Earth.

• **Do scientists today have any definite evidence that life exists on other planets?** (No.)

• **What leads them to believe that there might be life on other planets?** (Accept all answers.)

Explain to students that so far no evidence has been found to support the idea that life exists elsewhere in our solar system. Scientists have reason to believe, however, that life might exist elsewhere in our universe. This idea is supported by the fact that there are billions of stars in the universe similar to our sun. It is quite likely that many of these stars are surrounded by systems of planets. Scientists believe that there is a narrow region around each of these stars called a life zone. In the life zone exist the kinds of conditions that would make life possible.

• **Where do you think the life zone is located around our sun?** (The orbit of

impossible for an exoskeleton to withstand so much stress.

"At any rate," she said, "the creature would need elephant-sized, rather than insect-sized, legs to support such a mass."

Before Haley could continue, Fernandez reminded her of an important difference between Vivarium and Earth. The surface gravity of Vivarium is only one quarter that of Earth. This would allow for the evolution of a creature with a more massive body and thinner legs.

"All right," said Haley, "but you've got to admit, such a bulky creature would need at least six legs for balance. We have to picture it making its way over the rocky, uneven terrain of Vivarium in near darkness."

"Why stop at six legs?" I said. "And why consider only legs?" Again the two looked at me expectantly.

"According to the data brought back by our satellite," I went on, "the craters and canyons of Vivarium are covered with a thin film of water at least part of the year. Perhaps the wet, slippery rocks of this planet require the creature to have suction cups instead of legs and feet. Perhaps the crea-

ture's body is covered with suction cups so it could move even if it rolled over."

By the time our discussion drew to a close, it was clear to me that we can only make educated guesses about evolution on other planets. The guesses, of course, would be based on our knowledge of the environment of the planet. We had turned the ET into a creature that was the size of an elephant and had a head and outer skeleton that were similar to those of an ant. It had suction cups covering its body, and the ability to sense changes in heat and radiation levels.

I left Haley and Fernandez arguing about the creature's probable lung capacity. I was glad that I had decided to delay the battle over the ET. As it was, our discussion had taken up the last 3 hours of the voyage.

Well, Pete, I'm afraid I must sign off now. I can hear the first landing rockets firing. I wanted to get all of our ET discussion recorded before touchdown. Vivarium may very well prove to be the testing ground for a great many of our hypotheses about evolution on other planets. If everything goes as I expect it will, my next letter should make for some very exciting reading.

GAZETTE ■ 103

Critical Thinking Questions

1. How might humans be different if they had developed on a planet radically different from Earth? (Encourage students to speculate freely.)

2. Why do you think the planet is named Vivarium? (Answers may vary. The word *vivarium means* "a container for keeping or raising plants and animals"; from the Latin vivus, meaning "alive.")

3. How would you feel about visiting a planet populated with living creatures? Can you think of any possible dangers or opportunities in such an adventure? (Accept all answers.)

---

planet Earth.)

• **Why does the theory of a life zone tend to discount the possibility of finding life elsewhere in our solar system?** (Earth occupies the sun's life zone, and it is only in this region that the conditions necessary for life can be found.)

Discuss with students some of the factors that make a planet suitable for life. Some of these factors include atmospheric gases such as oxygen and carbon dioxide, as well as the absence of poisonous

gases; a moderate temperature; and the presence of liquid water.

• **According to this article, which of these conditions are present on the planet Vivarium?** (Abundant oxygen, lakes filled with water.)

• **Can you infer that Vivarium probably has a moderate temperature? If so, how?** (If the temperature were not moderately warm, water could not exist in the liquid state. The temperature also must not go above the boiling point of water, al-

though one cannot tell from the article whether the temperature there is hotter than on Earth.)

INDEPENDENT PRACTICE

▶ *Activity Book*

After students have read the Science Gazette article, you may want to hand out the reading skills worksheet based on the article in your *Activity Book*.

For Further Reading

If you have been intrigued by the concepts examined in this textbook, you may also be interested in the ways fellow thinkers—novelists, poets, essayists, as well as scientists—have imaginatively explored the same ideas.

Chapter 1: Earth's History in Fossils

Anker, Charlotte. *Last Night I Saw Andromeda.* New York: Henry Z. Walck, Inc.

Conrad, Pam. *My Daniel.* New York: Harper & Row.

Katz, Welwyn Wilton. *False Face.* New York: M. K. EcElderry Books.

Kelleher, Victor. *Baily's Bones.* New York: Dial Press.

Chapter 2: Changes in Living Things Over Time

Boulle, Pierre. *Planet of the Apes.* New York: Vanguard.

Dickinson, Peter. *Box of Nothing.* New York: Delacorte Press.

Lord, Bette Bao. *In the Year of the Boar and Jackie Robinson.* New York: Harper & Row.

Niven, Larry and Jerry Pournelle. *The Mote in God's Eye.* New York: Pocket Books.

Chapter 3: The Path to Modern Humans

Denzel, Justin. *Boy of the Painted Cave.* New York: Philomel Books.

Dyer, T. A. *A Way of His Own.* Boston, MA: Houghton Mifflin Co.

L'Engle, Madeleine. *Many Waters.* New York: Farrar, Straus, Giroux.

Millstead, Thomas. *Cave of the Moving Shadows.* New York: Dial Press.

Activity Bank

Welcome to the Activity Bank! This is an exciting and enjoyable part of your science textbook. By using the Activity Bank you will have the chance to make a variety of interesting and different observations about science. The best thing about the Activity Bank is that you and your classmates will become the detectives, and as with any investigation you will have to sort through information to find the truth. There will be many twists and turns along the way, some surprises and disappointments too. So always remember to keep an open mind, ask lots of questions, and have fun learning about science.

Activity Bank

COOPERATIVE LEARNING

Hands-on science activities, such as the ones in the Activity Bank, lend themselves well to cooperative learning techniques. The first step in setting up activities for cooperative learning is to divide the class into small groups of about 4 to 6 students. Next, assign roles to each member of the group. Possible roles include Principal Investigator, Materials Manager, Recorder/Reporter, Maintenance Director. The Principal Investigator directs all operations associated with the group activity, including checking the assignment, giving instructions to the group, making sure that the proper procedure is being followed, performing or delegating the steps of the activity, and asking questions of the teacher on behalf of the group. The Materials Manager obtains and dispenses all materials and equipment and is the only member of the group allowed to move around the classroom without special permission during the activity. The Recorder, or Reporter, collects information, certifies and records results, and reports results to the class. The Maintenance Director is responsible for cleanup and has the authority to assign other members of the group to assist. The Maintenance Director is also in charge of group safety.

For more information about specific roles and cooperative learning in general, refer to the article "Cooperative Learning and Science—The Perfect Match" on pages 70–75 in the *Teacher's Desk Reference*.

ESL/LEP STRATEGY

Activities such as the ones in the Activity Bank can be extremely helpful in teaching science concepts to LEP students — the direct observation of scientific phenomena and the deliberate manipulation of variables can transcend language barriers.

Some strategies for helping LEP students as they develop their English-language skills are listed below. Your school's English-to-Speakers-of-Other-Languages (ESOL) teacher will probably be able to make other concrete suggestions to fit the specific needs of the LEP students in your classroom.

• Assign a "buddy" who is proficient in English to each LEP student. The buddy need not be able to speak the LEP student's native language, but such ability can be helpful. (**Note:** *Instruct multilingual buddies to use the native language only when necessary, such as defining difficult terms or concepts. Students learn English, as all other languages, by using it.*) The buddy's job is to provide encouragement and assistance to the LEP student. Select buddies on the basis of personality as well as proficiency in science and English. If possible, match buddies and LEP students so that the LEP students can help their buddies in another academic area, such as math.

• If possible, do not put LEP students of the same nationality in a cooperative learning group.

• Have artistic students draw diagrams of each step of an activity for the LEP students.

You can read more about teaching science to LEP students in the article "Creating a Positive Learning Environment for Students with Limited English Proficiency," which is found on pages 86–87 in the *Teacher's Desk Reference*.

Activity Bank

WHERE ARE THEY?

BEFORE THE ACTIVITY

1. Gather all materials at least one day prior to the activity. You should have enough supplies to meet your class needs, assuming no more than six students per group.

2. Once the dots have been punched, you can save them from year to year in the plastic bags. Also, to save time, you can punch out all the dots prior to class. Be sure, however, to have a hole punch and extra paper on hand because students may need additional colored dots. You may use the same color and pattern paper or cloth for each group or have different colors and patterns per group.

PRE-ACTIVITY DISCUSSION

Before beginning the activity, review with students the concepts of natural selection and survival of the fittest and ask how they relate to the activity. Make sure students have read the entire activity before they begin. Prior to completing the activity, you may want to have students hypothesize the type of results they expect to observe.

TEACHING STRATEGY

Circulate through the room to make sure students have prepared their floral paper on a flat surface. Make sure students understand the need for the predator to look away prior to spreading the prey upon the floral surface.

DISCOVERY STRATEGIES

Discuss how the activity relates to the chapter by asking questions similar to the following.

• **How might the data you collected change if the floral pattern you used were replaced by a solid pattern?** (The data would almost certainly change, particularly if the solid pattern was the same color or shade as one set of colored dots.)

• **How might the ability of an organism to change its coloration to match its environment provide that organism with an added advantage for survival?** (The organism would be better able to hide from predators and would likely survive and reproduce better than organisms that could not change coloration.)

• **Aside from coloration, what other ways might an organism "hide" from its predators?** (Students will probably mention mimicry, or the ability to appear similar to a part of the environment. Another form of mimicry that may be mentioned is when an organism resembles another organism that predators avoid.)

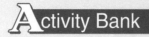

WHERE ARE THEY?

Natural selection is the survival and reproduction of those living things best adapted to their surroundings. To better understand how natural selection works, why not try this activity on camouflage, or the ability of living things to blend in with their background.

What Will You Need?

hole punch
colored construction paper (1 sheet of each of the following colors: black, blue, brown, green, orange, purple, red, white, yellow)
9 sealable plastic bags
80 cm × 80 cm piece of floral paper or cloth
transparent tape

What Will You Do?

1. Punch 10 dots of each color from the sheets of colored construction paper. Put the dots for each color in a different plastic bag.

2. Spread a piece of floral paper or cloth on a flat surface. Use transparent tape to attach each corner of the paper or cloth to the flat surface.

3. Choose one member of your group to be the recorder and another to be the predator. The other members of the group will be the prey.

4. Have the predator look away while the prey randomly spread the dots of each color over the paper.

WHAT WILL YOU SEE?

Make sure students have completed their entire data table. You may want to have the class complete another data table that reflects the observations of all the groups in the class.

5. Have the predator turn back to the paper and immediately pick up the first dot he or she sees.

Spreading the Dots

Picking up the Dots

6. Repeat steps 4 and 5 until a total of 10 dots have been picked up. Make sure that the predator looks away before a selection is made each time.

7. In a data table similar to the one shown, have the recorder write the total number of dots selected by the predator next to the appropriate color.

8. Have the recorder and the predator reverse roles. Repeat steps 4 through 7.

9. On posterboard, construct a data table similar to yours. Have your classmates record their results in this data table.

What Will You See?

DATA TABLE

Color of Dots	Number of Dots Selected
Black	
Blue	
Brown	
Green	
Orange	
Purple	
Red	
White	
Yellow	

What Will You Discover?

1. Which colored dots were picked up from the floral background?

2. Which colored dots, if any, were not picked up? Explain.

3. How did your results compare with your classmates' results?

4. If the colored dots represent food to a predator, what is the advantage of camouflage?

5. If the colored dots (prey) were to pass through several generations, what trends in survival of prey would you observe?

WHAT WILL YOU DISCOVER?

1. The colored dots that stand out against the floral background will be picked up.

2. The colored dots that blend into the background are the least likely to be selected because they are the least easily seen.

3. Student comparisons will depend on the colors present in the floral paper or cloth.

4. The predator will not see the food and the organism is less likely to be eaten.

5. The colored dots that blend into the background will increase in number, while the colored dots that stand out will decrease. Over time, only those colored dots that blend in will probably survive.

Activity Bank

BEFORE THE ACTIVITY

Gather all materials at least one day prior to the activity. You should have enough materials to meet your class needs, assuming no more than six students per group.

PRE-ACTIVITY DISCUSSION

Before beginning this activity, review the concepts of speciation and variation with the class. Remind students that variation is normal among a species and that natural selection may work for or against members of a species with a particular variation. Also remind students that evolution is often a slow process with minor variations occurring within an evolving species over time, but that sometimes rather rapid evolution, or punctuated equilibrium, can cause rapid (by evolutionary standards) changes in a species, often leading to new species.

TEACHING STRATEGY

Circulate through the room to make sure students follow all preparatory instructions accurately and that their initial setup is correct.

DISCOVERY STRATEGIES

Have students read over the entire activity before beginning. Then ask them to predict if this activity will be representative of gradual change or punctuated equilibrium (rapid change).

WHAT YOU WILL DISCOVER

1. Slight changes should exist in each evolutionary tree when compared with the rows above and below it.
2. Evolution, as described by Darwin, is the gradual change in a species over time. This activity illustrates evolution by showing how slight changes in the shape of the square can gradually produce new shapes.
3. The evolutionary tree of individual students will vary. They should, however, illustrate a gradual change in the shape of the square so that new shapes develop.

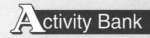

The fossil record shows that living things have evolved, or changed over time. How do these changes produce complex living things from simple ones? How can one group of living things evolve into many different groups? How can you show these changes in the form of an evolutionary tree (diagram that shows the evolutionary relationships among different groups of living things)? To find out the answers to these questions, try this activity. You will need the following materials: sheets of construction paper (red, blue, green, and black), metric ruler, scissors, posterboard, glue, pencil, compass.

What You Will Do

1. With the scissors, cut out 12 4-cm squares from a sheet of green construction paper.

2. Cut out one 4-cm square from a sheet of black construction paper. Then cut the square in half diagonally so that you

Construction paper

have two black triangles. Put one triangle aside for now and discard the other.

3. Repeat step 2 using a sheet of red construction paper.

4. With a compass, draw a circle that has a diameter of 4 cm on a sheet of blue construction paper. Put the blue circle aside for now.

5. Place the posterboard vertically on a flat surface. Draw a very faint line down the center of the posterboard.

6. Place one green square in the middle of the left side of the posterboard, about 5 cm from the bottom. Glue the green square in place.

7. Arrange 10 of the remaining green squares on the posterboard exactly as shown in the diagram on the left on page F109. You should have five rows of green squares: 1 square in the first row, 2 squares in the second and third rows, 3 squares in the fourth and fifth rows. Glue the squares in place.

8. Draw the arrows in as shown.

9. Go to the fifth row of green squares. Place the blue circle on top of the first green square so that it covers the square. Glue the blue circle in place.

10. On top of the middle green square in the fifth row, place the red triangle as shown in the diagram on the right on page F109. Glue the red triangle in place.

11. Above the third green square in the fifth row, place the last remaining green square so that you form a rectangle 8 cm × 4 cm. Then place the black triangle on top of the newly added fourth square. Glue the black triangle in place.

GOING FURTHER

If students complete the Going Further assignment, their illustrations should show a gradual change in organisms over time. This particular extension of the activity will be most interesting to artistic students, but you should suggest that all students complete the Going Further activity, regardless of their artistic abilities.

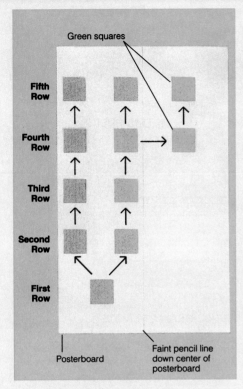

Green squares

Fifth Row	■	■	■
Fourth Row	■	■	■
Third Row	■	■	
Second Row	■	■	
First Row		■	

Posterboard

Faint pencil line down center of posterboard

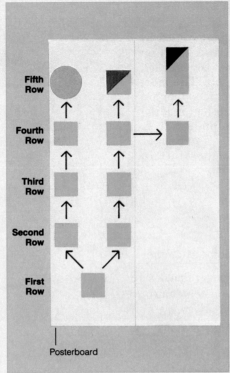

Posterboard

12. Observe the first row and the last row. Using the remaining sheets of colored construction paper, cut out the shapes that you think are needed to show the gradual change in shapes between the first row and the last row.

What You Will Discover

1. How does each row in your evolutionary tree compare with the row below it? With the row above it?

2. What relationship does this activity have with evolution?

3. Compare your evolutionary tree with those of your classmates. Are they the same? Are they different? Explain your answer.

Going Further

Replace the shapes in this activity with drawings of living things.

The metric system of measurement is used by scientists throughout the world. It is based on units of ten. Each unit is ten times larger or ten times smaller than the next unit. The most commonly used units of the metric system are given below. After you have finished reading about the metric system, try to put it to use. How tall are you in metrics? What is your mass? What is your normal body temperature in degrees Celsius?

Commonly Used Metric Units

Length   The distance from one point to another

meter (m)   A meter is slightly longer than a yard.
            1 meter = 1000 millimeters (mm)
            1 meter = 100 centimeters (cm)
            1000 meters = 1 kilometer (km)

Volume   The amount of space an object takes up

liter (L)   A liter is slightly more than a quart.
            1 liter = 1000 milliliters (mL)

Mass   The amount of matter in an object

gram (g)   A gram has a mass equal to about one paper clip.

            1000 grams = 1 kilogram (kg)

Temperature   The measure of hotness or coldness

degrees
Celsius (°C)   0°C = freezing point of water
            100°C = boiling point of water

Metric–English Equivalents

2.54 centimeters (cm) = 1 inch (in.)
1 meter (m) = 39.37 inches (in.)
1 kilometer (km) = 0.62 miles (mi)
1 liter (L) = 1.06 quarts (qt)
250 milliliters (mL) = 1 cup (c)
1 kilogram (kg) = 2.2 pounds (lb)
28.3 grams (g) = 1 ounce (oz)
°C = 5/9 x (°F – 32)

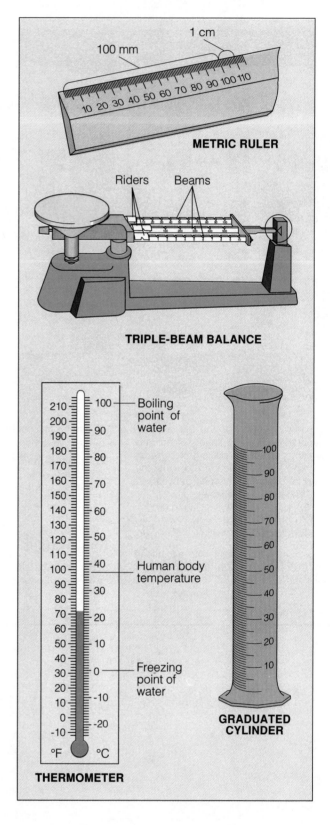

METRIC RULER

TRIPLE-BEAM BALANCE

THERMOMETER

GRADUATED CYLINDER

Glassware Safety

1. Whenever you see this symbol, you will know that you are working with glassware that can easily be broken. Take particular care to handle such glassware safely. And never use broken or chipped glassware.
2. Never heat glassware that is not thoroughly dry. Never pick up any glassware unless you are sure it is not hot. If it is hot, use heat-resistant gloves.
3. Always clean glassware thoroughly before putting it away.

Fire Safety

1. Whenever you see this symbol, you will know that you are working with fire. Never use any source of fire without wearing safety goggles.
2. Never heat anything—particularly chemicals—unless instructed to do so.
3. Never heat anything in a closed container.
4. Never reach across a flame.
5. Always use a clamp, tongs, or heat-resistant gloves to handle hot objects.
6. Always maintain a clean work area, particularly when using a flame.

Heat Safety

Whenever you see this symbol, you will know that you should put on heat-resistant gloves to avoid burning your hands.

Chemical Safety

1. Whenever you see this symbol, you will know that you are working with chemicals that could be hazardous.
2. Never smell any chemical directly from its container. Always use your hand to waft some of the odors from the top of the container toward your nose—and only when instructed to do so.
3. Never mix chemicals unless instructed to do so.
4. Never touch or taste any chemical unless instructed to do so.
5. Keep all lids closed when chemicals are not in use. Dispose of all chemicals as instructed by your teacher.

6. Immediately rinse with water any chemicals, particularly acids, that get on your skin and clothes. Then notify your teacher.

Eye and Face Safety

1. Whenever you see this symbol, you will know that you are performing an experiment in which you must take precautions to protect your eyes and face by wearing safety goggles.
2. When you are heating a test tube or bottle, always point it away from you and others. Chemicals can splash or boil out of a heated test tube.

Sharp Instrument Safety

1. Whenever you see this symbol, you will know that you are working with a sharp instrument.
2. Always use single-edged razors; double-edged razors are too dangerous.
3. Handle any sharp instrument with extreme care. Never cut any material toward you; always cut away from you.
4. Immediately notify your teacher if your skin is cut.

Electrical Safety

1. Whenever you see this symbol, you will know that you are using electricity in the laboratory.
2. Never use long extension cords to plug in any electrical device. Do not plug too many appliances into one socket or you may overload the socket and cause a fire.
3. Never touch an electrical appliance or outlet with wet hands.

Animal Safety

1. Whenever you see this symbol, you will know that you are working with live animals.
2. Do not cause pain, discomfort, or injury to an animal.
3. Follow your teacher's directions when handling animals. Wash your hands thoroughly after handling animals or their cages.

One of the first things a scientist learns is that working in the laboratory can be an exciting experience. But the laboratory can also be quite dangerous if proper safety rules are not followed at all times. To prepare yourself for a safe year in the laboratory, read over the following safety rules. Then read them a second time. Make sure you understand each rule. If you do not, ask your teacher to explain any rules you are unsure of.

Dress Code

1. Many materials in the laboratory can cause eye injury. To protect yourself from possible injury, wear safety goggles whenever you are working with chemicals, burners, or any substance that might get into your eyes. Never wear contact lenses in the laboratory.

2. Wear a laboratory apron or coat whenever you are working with chemicals or heated substances.

3. Tie back long hair to keep it away from any chemicals, burners and candles, or other laboratory equipment.

4. Remove or tie back any article of clothing or jewelry that can hang down and touch chemicals and flames.

General Safety Rules

5. Read all directions for an experiment several times. Follow the directions exactly as they are written. If you are in doubt about any part of the experiment, ask your teacher for assistance.

6. Never perform activities that are not authorized by your teacher. Obtain permission before "experimenting" on your own.

7. Never handle any equipment unless you have specific permission.

8. Take extreme care not to spill any material in the laboratory. If a spill occurs, immediately ask your teacher about the proper cleanup procedure. Never simply pour chemicals or other substances into the sink or trash container.

9. Never eat in the laboratory.

10. Wash your hands before and after each experiment.

First Aid

11. Immediately report all accidents, no matter how minor, to your teacher.

12. Learn what to do in case of specific accidents, such as getting acid in your eyes or on your skin. (Rinse acids from your body with lots of water.)

13. Become aware of the location of the first-aid kit. But your teacher should administer any required first aid due to injury. Or your teacher may send you to the school nurse or call a physician.

14. Know where and how to report an accident or fire. Find out the location of the fire extinguisher, phone, and fire alarm. Keep a list of important phone numbers—such as the fire department and the school nurse—near the phone. Immediately report any fires to your teacher.

Heating and Fire Safety

15. Again, never use a heat source, such as a candle or burner, without wearing safety goggles.

16. Never heat a chemical you are not instructed to heat. A chemical that is harmless when cool may be dangerous when heated.

17. Maintain a clean work area and keep all materials away from flames.

18. Never reach across a flame.

19. Make sure you know how to light a Bunsen burner. (Your teacher will demonstrate the proper procedure for lighting a burner.) If the flame leaps out of a burner toward you, immediately turn off the gas. Do not touch the burner. It may be hot. And never leave a lighted burner unattended!

20. When heating a test tube or bottle, always point it away from you and others. Chemicals can splash or boil out of a heated test tube.

21. Never heat a liquid in a closed container. The expanding gases produced may blow the container apart, injuring you or others.

22. Before picking up a container that has been heated, first hold the back of your hand near it. If you can feel the heat on the back of your hand, the container may be too hot to handle. Use a clamp or tongs when handling hot containers.

Using Chemicals Safely

23. Never mix chemicals for the "fun of it." You might produce a dangerous, possibly explosive substance.

24. Never touch, taste, or smell a chemical unless you are instructed by your teacher to do so. Many chemicals are poisonous. If you are instructed to note the fumes in an experiment, gently wave your hand over the opening of a container and direct the fumes toward your nose. Do not inhale the fumes directly from the container.

25. Use only those chemicals needed in the activity. Keep all lids closed when a chemical is not being used. Notify your teacher whenever chemicals are spilled.

26. Dispose of all chemicals as instructed by your teacher. To avoid contamination, never return chemicals to their original containers.

27. Be extra careful when working with acids or bases. Pour such chemicals over the sink, not over your workbench.

28. When diluting an acid, pour the acid into water. Never pour water into an acid.

29. Immediately rinse with water any acids that get on your skin or clothing. Then notify your teacher of any acid spill.

Using Glassware Safely

30. Never force glass tubing into a rubber stopper. A turning motion and lubricant will be helpful when inserting glass tubing into rubber stoppers or rubber tubing. Your teacher will demonstrate the proper way to insert glass tubing.

31. Never heat glassware that is not thoroughly dry. Use a wire screen to protect glassware from any flame.

32. Keep in mind that hot glassware will not appear hot. Never pick up glassware without first checking to see if it is hot. See #22.

33. If you are instructed to cut glass tubing, fire-polish the ends immediately to remove sharp edges.

34. Never use broken or chipped glassware. If glassware breaks, notify your teacher and dispose of the glassware in the proper trash container.

35. Never eat or drink from laboratory glassware. Thoroughly clean glassware before putting it away.

Using Sharp Instruments

36. Handle scalpels or razor blades with extreme care. Never cut material toward you; cut away from you.

37. Immediately notify your teacher if you cut your skin when working in the laboratory.

Animal Safety

38. No experiments that will cause pain, discomfort, or harm to mammals, birds, reptiles, fishes, and amphibians should be done in the classroom or at home.

39. Animals should be handled only if necessary. If an animal is excited or frightened, pregnant, feeding, or with its young, special handling is required.

40. Your teacher will instruct you as to how to handle each animal species that may be brought into the classroom.

41. Clean your hands thoroughly after handling animals or the cage containing animals.

End-of-Experiment Rules

42. After an experiment has been completed, clean up your work area and return all equipment to its proper place.

43. Wash your hands after every experiment.

44. Turn off all burners before leaving the laboratory. Check that the gas line leading to the burner is off as well.

Glossary

adaptation: change that increases an organism's chances of survival

adaptive radiation: process by which a species evolves into several species, each of which fills a different niche

cast: fossil in which the space left behind in a rock by a dissolved organism has filled, showing the same shape of the organism

Cro-Magnon (kroh-MAG-nuhn): early human that lived 40,000 years ago; *Homo sapiens sapiens*

evolve: to change over time

evolution: change in species over time

extrusion: igneous rock formation that forms on the Earth's surface

fault: break or crack along which rock moves

fossil: remains or evidence of a living thing

half-life: amount of time it takes for one half of the atoms of a sample of a radioactive element to decay

homologous (hoh-MAH-luh-guhs) **structure:** structure that evolves from similar body parts

imprint: fossil formed when a thin object leaves an impression in soft mud, which hardens

index fossil: fossil of an organism that existed on Earth for only a short period of time and that can be used by scientists to determine the relative age of a rock

intrusion: irregular formation of intrusive rock formed by magma beneath the Earth's crust

law of superposition: law that states that in undisturbed sedimentary rocks each layer is older than the one above it and younger than the one below it

mold: fossil formed in a rock by a dissolved organism that leaves an empty space, showing its outward shape

molecular clock: scale used to estimate the rate of change in proteins over time

natural selection: survival and reproduction of those organisms best adapted to their surroundings

Neanderthal: early human that lived 150,000 years ago to about 35,000 years ago; *Homo neanderthalensis* (nee-AN-der-thawl-ehn-sihs)

niche: combination of an organism's needs and its habitat

petrification: process by which once-living material is replaced by minerals, turning it into stone

primate: member of a group of animals that includes humans, monkeys, and about 200 other species of living things

punctuated equilibrium: periods in Earth's history in which many adaptive radiations occur in a relatively short period of time

sediment: small pieces of rock, shell, and other material that are broken down over time

trace fossil: mark or evidence of the activities of an organism

unconformity: eroded rock surface, pushed up from deeper within the Earth, that is much older than the new rock layers above it

Index